THE SCIENCE OF HISTORY IN VICTORIAN BRITAIN: MAKING THE PAST SPEAK

Science and Culture in the Nineteenth Century

Series Editor: Bernard Lightman

Titles in this Series

1 Styles of Reasoning in the British Life Sciences: Shared Assumptions, 1820–1858
James Elwick

2 Recreating Newton: Newtonian Biography and the Making of Nineteenth-Century History of Science
Rebekah Higgitt

3 The Transit of Venus Enterprise in Victorian Britain
Jessica Ratcliff

4 Science and Eccentricity: Collecting, Writing and Performing Science for Early Nineteenth-Century Audiences
Victoria Carroll

5 Typhoid in Uppingham: Analysis of a Victorian Town and School in Crisis, 1875–1877
Nigel Richardson

6 Medicine and Modernism: A Biography of Sir Henry Head
L. S. Jacyna

7 Domesticating Electricity: Expertise, Uncertainty and Gender, 1880–1914
Graeme Gooday

8 James Watt, Chemist: Understanding the Origins of the Steam Age
David Philip Miller

9 Natural History Societies and Civic Culture in Victorian Scotland
Diarmid A. Finnegan

10 Communities of Science in Nineteenth-Century Ireland
Juliana Adelman

11 Regionalizing Science: Placing Knowledges in Victorian England
Simon Naylor

FORTHCOMING TITLES

Communicating Physics: The Production, Circulation and Appropriation of Ganot's Textbooks in France and England, 1851–1887
Josep Simon

The British Arboretum: Trees, Science and Culture in the Nineteenth Century
Paul A. Elliott, Charles Watkins and Stephen Daniels

THE SCIENCE OF HISTORY IN VICTORIAN BRITAIN: MAKING THE PAST SPEAK

BY

Ian Hesketh

Published by the University of Pittsburgh Press, Pittsburgh, Pa., 15260
Copyright © 2020, University of Pittsburgh Press
All rights reserved

Cataloging-in-Publication is available from the British Library

ISBN 13: 978-0-8229-6636-4
ISBN 10: 0-8229-6636-0

CONTENTS

Acknowledgements	ix
Introduction: That Never-Ending Battle	1
1 The Enlarging Horizon: Henry Thomas Buckle's Science of History	13
2 The Sciences of History	35
3 Controversial Boys	55
4 Discipline and Disease; or, the Boundary Work of Scientific History	73
5 History from Nowhere	95
6 Broad Shadows and Little Histories	115
7 The Death of the Historian	133
Epilogue: Froude's Revenge	153
Notes	165
Works Cited	199
Index	219

ACKNOWLEDGEMENTS

This book was originally a PhD dissertation that was completely rewritten at least once and revised countless times before appearing in its present form. I was lucky enough to receive much encouragement and support throughout the research, writing, rewriting and revising process and as a result have many to thank. I must first of all acknowledge SSHRC doctoral and post-doctoral research fellowships that provided the necessary funds without which this project could not have been completed. I was also extremely fortunate to work as Assistant Editor and Project Manager for *The Oxford History of Historical Writing* for the last two years. This project not only renewed my interest in the manuscript, but the general editor, Daniel Woolf, was kind enough to give me the flexibility to work on it when needed. Working for Sandra Whitworth and the *International Feminist Journal of Politics* as well as for Bernard Lightman, Ian Slater and *Isis* during an earlier period in this book's incarnation was also extremely beneficial.

I was very lucky to have a PhD committee made up of Marlene Shore (who was my supervisor), Stephen Brooke and Bernard Lightman. Marlene kept me focused by forcing me to consider the appropriate questions when I wanted to head off in far too many directions; Stephen kept me honest when it came to British culture and politics; and Bernie helped me consider the vicissitudes of Victorian science and religion. They were also all very timely with their readings of the dissertation which helped speed the process along, particularly when a postdoctoral fellowship came calling. I was lucky, too, to have Christopher Green and Katherine Anderson ask me penetrating questions at my defence. Leslie Howsam, my external examiner, also gave me much to think about in her critique. Her work on British historical writing was an important resource that I relied upon for the dissertation while her recently published *Past into Print* was profoundly influential as I revised the manuscript.

In terms of research, Alvan Bergman arranged for my viewing of Henry Thomas Buckle's papers at the University of Illinois in Urbana-Champaign. His knowledge of the collection was very helpful and he was kind enough to share with me some of his immense knowledge of the Huth family. I would also like to thank John Reynolds at the University Library Cambridge; Elaine Pordes at the

British Library; John Hodgson at John Rylands University Library, Manchester; Richard Temple at the Senate House Library, University of London; and Colin Harris at the New Bodleian Library, Oxford.

History and Theory must be thanked for granting permission for the republication of sections of my article 'Diagnosing Froude's Disease: Boundary Work and the Discipline of History in Late-Victorian Britain' (2008), which appear in Chapters 3 and 4 and in the Epilogue. For the most part, the chapters are loosely based on papers that were presented at the Canadian Society of Church History (2008; 2010), the Canadian Society for the History and Philosophy of Science (2007), the History of Science Society (2006); the Canadian Historical Association (2006); and the New Frontiers in Graduate History Conference at York University (2005; 2002). I would also like to thank Claire Halstead and the Huron History Society, Ishita Pande and the Queen's University History Department Seminar Series, and Luis Aguiar and the UBC Okanagan Seminar Series for inviting me to give talks based on my research that allowed me to rethink much of my central arguments while I was in the process of revising. I also received great feedback about a previous version of Chapter 7 that was discussed by the Southern Ontario British History Reading Group in Toronto in 2005.

Many friends and colleagues have read parts of the manuscript or offered helpful suggestions during talks and discussions. I would like to especially thank Sahadeo Basdeo, Sandra Beardsall, Duncan Bell, Michael Bentley, Colin Duncan, Tod Duncan, Erika Dyck, Christine Grandy, Stephen Heathorn, Robert Holton, William Kelley, Ben Lander, David Leeson, Oliver Lovesey, Juan Maiguashca, Stuart Macintyre, Courtney Matthews, Ian McKay, James Morton, Liza Piper, Sandra den Otter, Geoff Read, Jim Robinson, Peter Russell, Ian Slater, Chuck Smith, Joseph Tawfik, Duane Thompson, Sarah Waurechen, Todd Webb, Sandra Whitworth, Maurice Williams, and Piotr Wrzesniewski.

I am also very grateful to John Beatty for helping me grasp Leslie Stephen's critique of Henry Thomas Buckle and the many varieties of the inductive method. Eric Nellis read much of the manuscript as it was being prepared for referees and he gave excellent advice on broader historical relevance. I greatly miss our regular meetings at the graduate pub at UBC. Daniel Woolf was kind enough to take the time out of his extremely busy schedule as Principal of Queen's University to read the entire manuscript in its close-to-final form providing extremely valuable critique. His ability to focus on the very broad issues while paying attention to the minute is extremely impressive and of course very helpful. James Hull has critiqued many versions of the manuscript, from when it was simply a few scattered thoughts in a proposal form to its current version. I'm not sure what I would have done without his friendship and continuous help over the years. Bernie Lightman has also helped shape this project since I was a Master's student in his Victorian science class. I am extremely grateful for his suggestion that I submit a revised version of the manuscript to Pickering & Chatto. He also provided many excellent suggestions for revision and he did a wonderful job as series editor for the manuscript. He found two excellent anonymous readers who gave many thought-provoking suggestions for revision. The manuscript has benefitted greatly because of that. The efforts of this manuscript's many readers have saved me from making countless mistakes. Any remaining mistakes are of course entirely my own.

Finally, I must thank my family for giving me the support necessary to pursue the charmed life of an academic. My parents, Bob Hesketh and Chris Turner, have provided endless emotional and financial support and I know they sacrificed a lot for me over the years. I must also acknowledge the wonderful support I have received from my step-parents, Anna Sorban and Jim Kemball, and my in-laws, Randy and Shelly Griffin. Kendra Hesketh, my sister, has always been encouraging. Both she and my mom were kind enough to listen to a summary of my book during a long drive between Vernon and Rossland, BC. They offered many important questions that had not even occurred to me. Cleo Griffin, my muse, was wonderfully patient with me during the research, writing, and revising of the manuscript. She must be sick of hearing about Buckle and Froude but I know that I'm extremely lucky that she found it amusing rather than irritating to discover countless scraps of paper with Lord Acton's name scrawled upon them and always asked before throwing them away. I am so very lucky to have her love and encouragement and I am excited to find out what the future holds for our new family.

... a historian is seen at his best when he does not appear.
—Lord Acton, 'The Study of History',
11 June 1895, Cambridge

We might as easily escape from our shadows.
—James Anthony Froude, 'Inaugural Lecture',
26 Oct. 1892, Oxford

INTRODUCTION: THAT NEVER-ENDING BATTLE

In September of 1854, the future popular historian Henry Thomas Buckle wrote to his friend Maria Grey, who was then becoming a powerful advocate of female education, to thank her for returning his much cherished six-volume set of August Comte's *Cours de philosophie positive* (1830–42). He had earlier lent Gray the volumes along with a set of instructions both for reading the difficult work and also for providing for its proper care. He suggested reading the 'Exposition' of the first volume but then skipping the analysis of the physical sciences contained in the first three only to return to them after reading the very important, in his mind at least, later volumes on the social sciences. He pressed her to keep *Philosophie positive* always in a safe place when not reading it, perhaps keeping it in a cupboard, 'as on several grounds I value it very much, and I never leave it out at home'.[1] When she returned the volumes because she was going on a trip to the country, Buckle expressed regret that she did not take them with her. Such a trip would have been the perfect time to try and get through Comte's rather unpleasant prose, but perhaps more importantly, 'in the country one particularly needs some intellectual employment to prevent the mind from falling into those vacant raptures which the beauties of nature are apt to suggest'. As Buckle explained, it was important to balance out that 'old antagonism ... between science and art' because '*that* is a battle which will never be ended'.[2] Nor, according to Buckle, should it be.

For Buckle, there would always be an inherent tension between science and art; the key thing is finding the appropriate balance between the two and maintaining that never-ending tension. This was a central aspect of his two-volume *History of Civilization in England* (published in 1857 and 1861). In this work, Buckle took up Comte's call to apply the same scientific methods of the physical sciences to the study of the human world, in particular to the study of human history. While Buckle rejected Comte's Religion of Humanity, that is, his attempt to apply the scientific method to the ordering of society, he was profoundly impressed with Comte's determination to place the study of humanity on a scientific footing. He argued that if properly studied it would be found that human history is governed by laws in much the same way that the motions of the planets are subject to the laws of gravity. The immense popularity of Buckle's book made 'the science of history' a surprisingly popular topic of conversation among the chattering classes and, perhaps not coincidentally, making history a science became a widespread goal within the burgeoning historical profession.[3]

But Buckle's reasons for wanting to put history on a scientific footing were quite a bit different from most other English proponents of scientific history, and he had a very different conception of just what it meant to make history a science. What Buckle found so unappealing about the contemporary state of historical writing was the failure of historians to generalize, to connect seemingly unconnected phenomena. He bemoaned the growing specialization of historians where one knows nothing about the economy while another knows nothing of society, the seeming lack of historical imagination and philosophical speculation, and, perhaps most importantly, the increasing devotion to facts and facts alone.[4] Buckle would have very much sympathized with Charles Dickens's lampooning of fictional historian and fact-grubber Thomas Gradgrind who teaches his students that 'Facts alone are wanted in life. Plant nothing else, and root out everything else.'[5] Historians, for Buckle, certainly needed facts but they needed to think about them more imaginatively and speculatively. They needed ideally to be more like Newton and let their imagination be carried off from the discovery of a single (or simple) fact like an apple falling from a tree to the nether regions of the universe.[6] Buckle advocated appropriating the new science of statistics while embracing a poetic imagination that would allow the historian to uncover history's equivalent to physics' gravity. A true science of history for Buckle could only be possible with an equal acceptance of the necessary artistic elements inherent in scientific discovery and dissemination, an acceptance, in other words, in that necessary and ever-present tension between science and art.[7]

Those many English historians who followed Buckle in calling history a science were less accepting, however, of what he believed was an inherent but necessary tension between science and art and they were less convinced by his particular brand of scientific history.[8] For this group, the problem with the discipline of history was not that historians had yet to discover laws of historical development, but that there seemed to be very little to distinguish history from other forms of literature, particularly that of fiction. The first half of the nineteenth century witnessed a general blurring of the boundaries between the historical and the fictional. This blurring was most particularly felt in the form of Romantic historical novels that often purported to give a true sense of past historical times and persons without having to rely on the constricting nature of mundane and minute facts that would generally turn such interesting stories into a form of writing that was, according to Walter Scott, 'dryasdust'.[9] Just as a revolution in printing and literacy was taking place,[10] historical novels flooded the marketplace, presenting a picture of the past that was romantic and exciting, poetic and heroic, and very often published under the explicit label of 'history' as was William Makepeace Thackeray's novel *The History of Henry Esmond* (1852). Literature about the past became hugely popular, and many historians of the first half of the nineteenth century seemed to generally embrace what Ann Rigney calls a 'romantic historicism' that accepted the sublime nature of the past, of its central unknowability, while appropriating rather than rejecting the literary and narrative techniques of their fictional counterparts. Men of letters such as Thomas Carlyle and Thomas Babington Macaulay wrote works of history that very much read like Romantic novels. Carlyle, in particular, was not afraid of blending the fictional with the factual, while Macaulay wrote about moments in time that read like scenes right out of Scott's Waverly novels.[11] There was very little separating the novelist from the historian and history as the subject of analysis was certainly not limited to a select group of historical observers.

Most historians who embraced the science of history in the second half of the century wanted to transform the public perception that, according to Carlyle, 'all men are historians'.[12] For these historians, making history a science meant making history above all an autonomous discipline of study, where historians would have to adhere to a common set of methodological assumptions thereby separating the *trained* historian from a mere man of letters. As we will see in Chapter 2, this group, led by the Anglicans William Stubbs, J. R. Seeley and Edward A. Freeman, and the liberal Catholic Lord Acton, looked not to post-Revolutionary France for their scientific inspiration (as was often pejoratively claimed of Buckle) but to Protestant Germany, finding in the work of Leopold von Ranke in particular a scientific model worth replicating.[13] This was a method that emphasized the importance of individual facts that were to be uncovered not by reading other historians but in the newly opened and expanding state archives. Ranke also stressed that the presentation of such facts must be devoid of an individual subjectivity and literary style, that the facts alone must be presented. The historian, for Ranke, was no longer a literary genius writing about the past in the same way a novelist writes about the fictional world, full of judgements and lessons of morality. The true scientific historian merely presents the past as it actually happened – '*wie es eigentlich gewesen*'.[14]

This was a message that accorded more closely with most English historians' religious beliefs and it actually seemed to be the application of an English tradition of science stretching back to Francis Bacon and Robert Boyle.[15] What is more, the royal road to science seemed to follow the general adoption of the Baconian method as was clearly the case with new sciences such as geology in the early nineteenth century. Baconianism was becoming an 'idol of the marketplace', and historians looking to make their own discipline a science found in Ranke's work a powerful exemplar of Baconian induction applied to historical phenomena.[16] Ranke's inductive approach was, therefore, an authoritative scientific method that English historians had a much easier time embracing than Buckle's rather deterministic model that was also still explicitly literary. Indeed, Buckle's method seemed to give prominence to generalists or literary geniuses, whereas the Rankean method promoted the creation of historical workers and researchers, those ready to engage in a difficult and gruelling analysis of specialized sources and subject matters that would be presented to the public in as accurate a form as possible. They wanted in particular to get away from the Romantic portrayals of English history that tended to skew the boundaries between history and literature, science and art.

Of course, there were still many historical writers in the second half of the century who felt little allegiance to any 'science of history' and they set about carrying on the English tradition of Romantic historicism and maintaining the inherent connections between history and literature. Chapter 3 considers the two most outspoken artistic historians in this period, Charles Kingsley and James Anthony Froude, two fairly radical Anglicans who were both highly influenced by Thomas Carlyle and who also had earlier in their literary careers written quite controversial works of autobiography that masqueraded as fiction. They were both openly sceptical of Buckle's science of history which, they argued, undermined the essential indeterminacy of the past. Froude in particular openly doubted whether history could be a science at all, even under the Rankean approach.[17] The historian was first and foremost a writer in Froude's mind and he became a powerful and popular voice upholding history's art against its science throughout the later Victorian period.

In order to establish the inductive scientific method of history as the one true historical method, historians such as Stubbs set out to establish a clean break between their scientific and serious pursuit and the more artistic and immature practice of their literary enemies. Romantic history was presented as a practice of a time when the actual facts of the past were little known and the historian had to generalize and imagine as much as possible in order to establish some sense of what happened. Such studies were by necessity subjective and overly burdened by the presence of the author. The state of historical knowledge in the nineteenth century, however, rendered Romantic historical writing an antiquated practice that must be abandoned. In the place of the subjective, untrustworthy, and inherently controversial writer of *belles lettres* only interested in entertaining and selling books with little regard for the actuality of the past, scientific historians advanced an objective, trustworthy, and disinterested historical worker only interested in presenting the past as it actually happened. For Stubbs, in particular, the historian was no literary genius but an archival worker tasked with uncovering and disseminating the facts of the past.[18] Or, as Lord Acton would have it, it was not genius that made the historian but the inductive method.[19] This was a message that the historical community would for the most part embrace, though, as we will see, even among the Rankeans there was by no means absolute consensus on just what that message meant.

What the Rankean historians did agree upon, however, was the necessity of ridding the discipline of what Freeman would refer to as the 'torrent of Lives' that had been flooding the general reading market, those Romantic studies of great men and women masquerading as serious history.[20] Part of the process of establishing the Rankean approach as the orthodox historical methodology involved emphasizing the *discipline* of history in establishing an appropriate demarcation between good history and bad, between a proper scientific analysis of the past and a mere Romantic narrative of a largely imaginary past. Here English historians were also following the lead of their more properly scientific colleagues who were seeking to establish more formal boundaries between the work done by men of science and that done by popularizers and other interlopers who often appropriated the work of others and at times even put forward such findings to suit their own interests while embracing a Romantic form of narration that would appeal to broader audiences.[21] Robert Chambers's hugely popular and anonymously published *Vestiges of the Natural History of Creation* (1844) is a case in point. That work sought to synthesize current work on the question of transmutation in order to explain the history of essentially everything, written in a pleasing Romantic narrative that found its way onto many Victorians' night tables and reading stands. In attempting to establish themselves as the proper investigators and disseminators of scientific knowledge, men of science such as Thomas Henry Huxley often viciously attacked the work of popularizers, supposed imposters and promoters of pseudo-science like that of phrenology. *Vestiges* was, in particular, brutally attacked and not just because it had popularized a heretical theory of transmutation. Huxley, for instance, was simply disgusted that a work of such shoddy scholarship could find a large readership simply because of the way in which it was written.[22] The boundaries that were established through fights such as this one, between scientific naturalists, natural theologians, North British physicists, popularizers of various stripes, gentlemanly naturalists and professionalizing men of science, were part of a broader struggle for cultural authority that historians of science have found so central in the formation of Victorian science.[23]

Scientific historians were very much a part of this struggle for cultural authority. Much like the burgeoning men of science, scientific historians criticized works that seemed to rely on an overblown Romantic narrative. Such reviews would essentially be foils to promote the Rankeans' supposedly more serious and scientific approach. Buckle, who was largely seen as a popularizer of pseudo-scientific ideas, suffered from such epistemic boundary work, as did Macaulay, Carlyle, and Kingsley. However, it was Froude who likely suffered the most sustained criticism, as we will see in Chapter 4.[24] Froude's studies were not only written in an artistic style but they were immensely popular and what is more, he relied quite extensively on archival research thereby undermining the chief complaint about Romantic histories, that they were not based on an extensive reading of the primary sources. Not only was Froude suggesting that artistic history could be based on primary research, his historical studies also made it more than clear that such work need not be unreadable but could be enjoyed by the general public.

The chief disciplinarians for the Rankean method were Freeman and Seeley who did much to promote the new scientific method while challenging the merits of a history that seemed to rely more on artistic flair than a strict representation of the facts. Their primary concern was the way in which they believed artistic historians skewed the facts of the past in order to please a popular readership. How could such writers be true to the past while entertaining such a wide audience? Seeley, in particular, believed that such a thing was impossible and he sought to show that proper scientific history would simply be uninteresting for the general reader, that it was ideally for historical peers alone.[25] Freeman was not entirely convinced that good scientific history could not also be interesting, but he was sure that historians who explicitly embraced an artistic method simply could not tell the past as it actually happened. Because of their method, they would always be tempted to shape the facts to fit a more fascinating narrative.[26] He spent much of his career victimizing Froude, engaging in painstaking reviews that sought to point out the countless inaccuracies engendered by Froude's false historical method. There were very real consequences for not adhering to history's new scientific method. The diagnosing of 'Froude's disease' was one such consequence, that is, the artistic disease of inherent inaccuracy, which became a powerful stigma that would blight Froude's name well after his death.[27]

It was certainly much easier to explain what was wrong with artistic history than to explain just what a proper inductive science of history was supposed to look like. If both scientific and artistic historians admitted that histories had to rely on primary authorities, the difference suddenly becomes one of degree rather than one of kind. What scientific historians began to stress about Froude and others was not the lack of research but rather the subjective presence in their writing. Scientific historians, on the other hand, wanted ideally to absent themselves from their discourse and in this way become truly disinterested while letting the past speak for itself. In other words, their identity as scientific historians was connected to a particular form of objectivity, though 'disinterest', 'impartial', 'impersonal' or 'self-restraint' were the key terms more likely to be used by these historians. They sought to let the past speak by being true to their sources, that is by detaching their own subjective views from the study and dissemination of those sources. In this way Rankean historians seemed to mimic a 'mechanical' form of objectivity that has been noticed in the work of naturalists during the same period who sought to let nature speak for itself,[28] though it is important to note that it was not just a passive, self-effacing process – an escape from self. It was also an act of self-overcoming, of the creation of a new detached self who could be trusted to impart knowledge of the past while even making judgements about what happened.[29] There was something deeply moral that underpinned the creation of this detached scientific self as if it was the primary duty of the truly scientific historian. This was a duty that was clearly shirked by Romantic historians whose obvious literary style was a clear sign of an author's presence and therefore of his or her weak historical mind. At its most ideal, a true scientific history would be devoid of style and perspective, of all imprints of individuality and subjectivity, something that could only be the product of a strong-willed devotion to overcoming Romanticizing temptations. The scientific historian would essentially be a trusted mediator of historical truth, much like the men of science who sought to establish themselves as trusted authorities on natural knowledge.

Actually fulfilling the requirements of such a method is surely impossible, whether for the historian or the naturalist, but a few historical monographs were held up as powerful exemplars of just this method, most notably William Stubbs's *Constitutional History* (1873–8), a book where it was said that the author's 'power of restraint' holds any 'personal expression' in check 'until the last paragraph of the third volume' where his presence is finally felt.[30] Other publishing ventures such as the quarterly *English Historical Review*, first published in 1886, and the Acton-directed *Cambridge Modern History*, first published in 1902, also provided a space outside of the general periodical press for the scientific historian to write under the inductive methodology, free, in theory, of the pressures of the dramatic dictates of popular publishers and their reading audiences (see Chapter 5).

This is not to suggest, however, that those promoting scientific history at the expense of Romantic history wanted to make history as unreadable as possible, as a mere list of facts. While, rhetorically, some Rankean historians such as Seeley approached such extremes in their methodological pronouncements, they were never quite as 'dryasdust' as someone like Scott or Dickens might have lamented, and certainly many of the so-called Romantics shared much of the same respect for the importance of primary research and of historical facts, Froude being a primary example. Much of the battle between the two groups seemed to have less to do with the actual content of the histories than with the answer one might give to a fundamental question: Is history a science or a form of art? The Romantic Froude and the Rankean Seeley would have given knee-jerk responses to such a question, reflecting the opposite sides to this seemingly binary debate – and yet their work, much like most other historical writing of the period, reads as if underpinned by a more nuanced method. Indeed, Rankean historians who were adamant that history was a science, that it *had* to be a science, still wanted to make their work more palatable to a wider audience than their normative one of historical peers and they struggled with finding a balance between their new scientific demands and their personas as very public literary figures. In the same way that men of science such as Huxley sought to erect boundaries between their own work and that of interlopers even while seeking ultimately to diffuse their views about science to society at large, scientific historians above all wanted their work to be appreciated, their findings disseminated, and their method embraced in the classrooms as much as in the reading halls and in parliament.[31] Some historians, like Freeman, became very good at addressing different audiences with ultimately the same message, at wearing different hats depending on the occasion, and at writing very specialized and (admittedly) repetitive studies largely for historical peers and more popular studies for general audiences, while others, like Stubbs, seemed only capable of addressing an audience of peers alone, unable to change hats as it were.

There was also a general assumption by Rankean historians that there was a large reading audience that was not being served by the Romantic works that filled the bookshops, and that a wider audience could be found for more serious scholarly work. But convincing publishers of the value of their work was just as important as convincing the public and, as Leslie Howsam has shown in a series of important studies, historians often had to bend to the demands of their publishers rather than suffer the consequence of refusing to bend their scientific principles.[32] That consequence was silence, and every historian under study here, perhaps with the exception of Seeley, wanted a large public audience for their work (and Seeley found a large audience whether he wanted one or not). Some historians were certainly better able to speak to that audience than others, and that supposedly large readership clamouring for the normative Rankean work promoted by the likes of Freeman, Seeley, Acton and Stubbs, never quite materialized as was made most clear by the rather small group of readers who subscribed to the *English Historical Review* despite attempts to change its initially narrow editorial mandate (discussed in Chapter 6).

What is more, as discussed in Chapter 5, there were moments when the consensus among the Rankeans fell apart. Previously unspoken problems in the rhetorical method were exposed, such as when Mandell Creighton's devotion to a more 'mechanical' presentation of the facts of the Catholic Counter-Reformation did not do justice, according to Lord Acton, to the vast abuses of the Catholic Church. Creighton, of course, believed that he was following the Rankean prescription of avoiding making his history into a series of judgements on the past. Acton, who was seen as the foremost practitioner of the Rankean method, suddenly appeared as a great defender of an older view of history where the past was judged by the preconceptions of the present. Their fascinating debate about the role of judgements in historical analysis, as well as the necessary centrality of a universal Christian moral code, sheds much light on just how divided historians could be who appeared, on the surface at least, to agree on fundamental methodological principles.

The gap between the practice and principle of an inductive scientific history is considered further in Chapter 7 where the obituaries of the main Rankean historians are discussed as a forum where their historical careers received critical appraisal from a younger generation. The obituaries that mark their deaths read like death notices not just for the historians but for the scientific history they promoted. It is not as if the obituary writers did not appreciate the critical methods of research and writing that the Rankean historians promoted, to the contrary. However, the obituary writers all highlighted the artistry of their works, or regretted when it was not relied upon more fully. The art of history was not only alive and well at the end of the century, it was deemed an unspoken part of the scientific historian's method from the beginning, despite the deceased's methodological claims to the contrary. By the turn of the century, there seemed to be a general acceptance that a certain amount of poetic imagination along with a reliance on primary sources, makes for the best history, that history was at once a science and a form of art. It would be going too far to suggest that the tension and even battle between science and art in historical writing ends at the turn of the century as the boundaries separating the two as well as the grudging overlap between them has continued to be negotiated and renegotiated throughout the twentieth and twenty-first centuries much like it was during the Victorian period. But, in Britain at least, it has been over 100 years since it was fashionable to try and let the past speak for itself despite the long hegemony of empiricism during the same period.[33]

Indeed, it was not long after Lord Acton's death in 1902 that it became common currency not to take seriously the specific scientific claims typical of his generation. It is for partly this reason that the historiography on the topic has received scant attention outside of a few articles and chapters.[34] There has been much work done on Victorian history-writing from very general studies on the period as a whole to monographs on individual historians.[35] There is also a considerable historiography on the institutionalization of history.[36] The historians of English history in particular have been analysed and their studies are often grouped under a term that at the time had a strictly political meaning: whig.[37] Historians as methodologically diverse as Froude and Freeman, Macaulay and Stubbs are said to have shared the same essential narrative framework of English history, one that foregrounded tradition, continuity, progress and a natural historical process.[38] No doubt there are elements of this tradition that extend throughout many historical works that have been produced in the Victorian period and beyond. Indeed, these studies tell us a great deal about the content of Victorian history-writing, but they tell us little about the actual engagement historians were making with scientific methodologies when their scientific claims are dismissed as mere rhetorical strategy.

This book takes seriously the scientific discourses that were constructed throughout the period under study, whether from the pen of Henry Thomas Buckle or that of his more conservative inductive critics. The focus here is less on the content of the monographs and historical studies of the historians discussed and more so on those places where we find the historian engaged in methodological analysis. That was more typically found in prefaces, in lectures, in periodical articles, in reviews and in letters – those places where historians spoke specifically to one another while they at times also addressed the public. This leaves much of the analysis at the level of rhetoric but not only so. In considering the creation of scientific history in Victorian Britain we are essentially considering the creation of a community and the construction of its central myth. Such requires the consideration of a broader historical context, one that includes the biographies and personalities of the key figures as well as the religious and social debates of the day as they informed the much more narrow struggle to make history a science.

Given that this book is about the attempt to make history a science and establish a scientific historical method, a word, perhaps, should be said about the method that underpins this study, though such a statement in some ways undermines the narrative strategy employed throughout, one that seeks to analyse through description and illustration rather than through explicit causal interpretation. The more I study this period and the more in-depth my research becomes, from initially examining published sources to expanding my research base to include the correspondences and other archival and unpublished materials, the more difficult it is to make specific causal interpretations about what it all means. It is perhaps necessary that the more one studies about a given topic the less easy it becomes to make bald statements of generalized fact without countless exceptions and hedges, that the analysis must become more nuanced and the general argument less specific and determinate. It has become clear to me that, ironically, telling the story of 'what actually happened' is less and less possible the more one learns about it and I have become sceptical of the historian who is far too sure of him or herself, who is able to condense the vicissitudes of any event into a few choice statements, making it far too apparent that what happened happened necessarily in a certain way. Indeed, it should be clear to anyone who has taken the time to thoroughly investigate the past that 'replaying life's tape', as Stephen Jay Gould once so eloquently put it in a slightly different context, would bear witness to an entirely different narrative thanks to countless contingencies.[39] With this in mind I have sought, wherever possible, to let the facts speak for themselves but not as Lord Acton would have done had he written a book, but rather by letting the subjects speak in their own words as often as is possible and without the intrusion of a metahistorical voice telling the reader what it is all supposed to mean. I certainly have my own broad interpretation about what happened as the introduction will have made clear, and even more particularly what, for instance, was meant by a particular quoted statement. But the analysis throughout is meant to be subtle and presented in the organization of the material and in illustrations and descriptions; it is an analysis that should allow the reader to make his or her own way. While this method has been influenced by much more recent studies of the contingency of historical events, of the strangeness and incomprehensibility of the past, and of the importance of self-conscious literary modes of representation in expressing such contingency,[40] it is not a method that would have been entirely foreign or incomprehensible to the diverse historians under study. A few might even have approved.[41]

1 THE ENLARGING HORIZON: HENRY THOMAS BUCKLE'S SCIENCE OF HISTORY

> The occurrences which contemporaries think to be of the greatest importance, and which in point of fact for a short time are so, invariably turn out in the long run to be the least important of all. They are like meteors which dazzle the vulgar by their brilliancy, and then pass away, leaving no mark behind.
>
> Henry Thomas Buckle, 'Mill on Liberty', *Fraser's Magazine* (1859)

On 19 March 1858 'the doors of the Royal Institution were opened some time before the usual hour to admit the throng of fashionable people who had collected', recalls Henry Thomas Buckle's (1821–62) friend and biographer Alfred Henry Huth, 'and by the usual time for opening the theatre was [sold-out] ... by a brilliant and excited audience'. Those lucky enough to get tickets 'were crammed from floor to ceiling' in order to be present for the first and, what would turn out to be, the only public lecture by Buckle who himself could not procure a sufficient number of tickets for his guests.[1] 'It is very hard that you should be limited because of your just popularity', John Barlow, the organizer of the event, wrote to Buckle. 'But what can be done? I can not expand the lecture-room, nor prevent members from exercising their right to indulge themselves and their friends with a high intellectual gratification.'[2] The members to whom Barlow referred belonged to the Royal Institution and they were hoping to get a glimpse of the 'great Buckle' while hearing him speak on a surprisingly popular topic: the science of history. Buckle would not disappoint. None other than the natural philosopher and polymath William Whewell (1794–1866) 'found everybody in London talking about Mr. Buckle's lecture'.[3]

Buckle had been catapulted into celebrity status thanks to his first volume of the *History of Civilization in England*, published by J. W. Parker in 1857, a work that was the sensation of the literary season.[4] It was provocative, topical, well-written and, what is more, it was shocking. Looking back on the period from the vantage point of 1880, Leslie Stephen (1832–1904) argued that there were 'two great intellectual shocks' that rocked the 'generation which was growing to maturity in the decade 1850–60'. One of the shocks, no surprise here, came by way of Charles Darwin's *Origin of Species* published in 1859. The other shock was provided by Buckle and his *History of Civilization*, preceding Darwin's by just two years. This would have been surprising for Stephen's readers because, as he explained, Buckle's influence, in contrast to Darwin's, 'has faded; his name is rarely cited by the eager disputants in the exciting controversies of the day'. And yet, Stephen explained, 'Buckle's performance perhaps seemed the most important at the moment'.[5]

Indeed, for the space of about two years it seemed as if Buckle's name would appear alongside the great historical geniuses who had previously shone a bright light over the course of history and he was at the height of his popularity and power as he prepared to dazzle the London literati with a display of his eloquent and grand historical speculations. But his lecture at the Royal Institution represented an absolute peak in his popularity and influence, only to be followed by a precipitous decline. Within the space of only a few years, Buckle would be dead of a fever at the age of forty, abandoned by his travelling companion just as he fell ill, and buried in a Protestant Damascus cemetery that would soon after be desecrated by radical Muslims, with nothing more than a broken cross to mark his final resting place.[6] Leaving this world behind with no distinctive mark, 'to pass away and make no sign',[7] was his worst fear, made literal by the unfortunate circumstances of his death, figurative by the fact that Buckle and his *History of Civilization* faded from public discourse almost as quickly as they seemed to take hold of it. But take hold of it Buckle did.

There was something truly quite shocking, according to Stephen, about Buckle's historical pronouncements and analysis that jolted the Victorian psyche while speaking to a few central and growing anxieties. At the book's core was a simple premise that the past, much like nature, is determined by laws. Buckle argued that it was the historian's primary duty to uncover those laws in order to truly explain how history developed, how nations, particularly, but not only, England, have progressed, and perhaps provide the reader in particular and society in general with a glimpse into the future in order to envision and properly plot out the future course of progress. Shocking in all of this was Buckle's assertions that God, at least directly, and morality played no role in shaping the course of history. History was determined by laws that knew no right or wrong and no amount of human intention, good or bad, could transcend those laws. To think that humans might be at the mercy of some perfunctory law and not their own free will or the guidance of a benevolent God, was both frightening and exciting. It also made the previously unknown Buckle the lion of the 1857 literary season at the age of thirty-six.

Buckle claimed to have spent the better part of his adult life researching and preparing to write a book that he envisioned as early as 1842.[8] By February 1853, he had been working on the project for several years and had a very clear idea about the argument and content of the soon to be completed manuscript. He wrote to the Lord Kintore, who had requested a summary of the work in progress, that

> I have been long convinced that the progress of every people is regulated by principles – or, as they are called, Laws – as regular and as certain as those which govern the physical world. To discover those laws is the object of my work. With a view to this, I propose to take a general survey of the moral, intellectual, and legislative peculiarities of the great countries of Europe, and I hope to point out the circumstances under which those peculiarities have arisen. This will lead to a perception of certain relations between the various stages through which each people have progressively passed. Of these general relations, I intend to make a particular application; and, by a careful analysis of England, show how they have regulated our civilization, and how the successive and apparently the arbitrary forms of our opinions, our literature, our laws, and our manners, have naturally grown out of their antecedents.

Buckle went on to say that the success of his plan would largely 'depend upon the fidelity with which I carry that scheme into execution, and on the success of my attempt to rescue history from the hands of annalists, chroniclers, and antiquaries'.[9] From fairly early on, then, Buckle had a clear idea, ambitious though it may have been, about what his *History* would look like and how it would, if successful, transform the writing of history into a science of the highest order.

Unfortunately Buckle's intellectual ambitions were often frustrated by his poor health. He was diagnosed at a young age as being 'sickly', and doctors warned of what might happen should the young boy overly tax his brain. Because of this he was able to avoid formal educational settings and was, for the most part, self-taught. For Leslie Stephen (and likely many others) it was the informal nature of Buckle's education that led to his very peculiar science of history; it gave him the freedom to pursue his own intellectual interests but in a haphazard way, depriving 'him of the main advantage of schools and universities – the frequent clashing with independent minds – which tests the thoroughness and solidity of a man's acquirements'.[10] Yet, if Buckle followed his own 'peculiar' interests he did so with a great amount of self-discipline in order to gain the knowledge to complete his project but also to do so in a way that would not jeopardize his health. This is made abundantly clear in his diary where his daily work activities were meticulously recorded.

> *Monday, November* 24, 1851, *Brighton.* – Rose at 8. Walked half an hour and then breakfasted. From 10.5 to 12 read German. From 12 to 1.30 read Mill's Analysis of Mind, i., 66–140. Walked one hour and a half, and from 3.40 to 4.30 made notes from Leigh Hunt's Autobiography. From 4.30 to 6.20 read Lord Lyttleton's Memoirs and Correspondence, i., 246, to vol. ii., p. 580 (the paging of the two volumes is continuous). Dined at 6.30. In bed at 10.20, and to 11.30 read Beattie's Campbell, ii., 61–236.[11]

Buckle's personal library numbered around 20,000 volumes by 1859, enabling the sickly man to avoid public libraries and diligently work toward completing his ambitious project on his own schedule.[12]

Buckle was able to afford such a massive library as well as the time necessary to engage in daily study because he came from a fairly wealthy merchant family. His father, Thomas Henry Buckle, was a partner in a shipping firm that traded in the West Indies. The firm, 'Buckle, Bagster and Buckle', did quite well though the plan was to bring the younger Buckle into the fold when he turned seventeen, a plan that worked out for only a short time. In 1840 Buckle's father slipped and broke his arm; his health was already suffering from consumption and he never recovered. He left his son £20,000 which was just enough to allow Buckle the freedom to avoid work the rest of his life. Buckle carefully budgeted his inheritance, and he found that there was plenty of money that could be spent on his growing book collection, as well as a few pounds that could be spent on travel.

After his father died Buckle travelled a great deal throughout the Continent and in the Middle East and his eyes were opened to a wide range of political ideas. He began to refer to himself as a radical and free thinker, thoroughly shedding the Tory and High Churchman beliefs that Buckle's father had instilled in him as a youth. He also made a concerted effort to learn the language that was spoken in the many countries he visited. He claimed that he could read eighteen foreign languages and could speak six by the time he was twenty-nine years old. His travels had a profound influence on the way in which he thought about history and society more generally. He was particularly predisposed to French social and political theory but he was also profoundly interested in the history and progress of English science. These two streams of thought played an important role in shaping his views of history.[13]

Buckle tended to align himself with slightly radical views of the day and this was certainly the case when it came to science. He was heavily critical of the popular Bridgewater Treatises, a series of scientific studies devoted to studying aspects of nature in order to uncover 'the Power, Wisdom, and Goodness of God as Manifested in Creation', as set out in the Earl of Bridgewater, Francis Henry Egerton's will, which had set aside £8,000 to be paid to the future authors. Not without cause, Buckle believed that the Bridgwater authors had more interest in proving God's existence than in studying nature to uncover its laws. He was particularly disappointed with William Whewell's volume because he clearly shared a great deal with the Cambridge Professor of Philosophy particularly concerning issues of method, but he could not forgive the man for seemingly putting religious views ahead of the strictly scientific. Rather, Buckle was attracted to the ideas of Anglican radicals like his friend Baden Powell who was an early advocate of evolution and a powerful voice in favour of biblical criticism as his essay in the much maligned *Essays and Reviews* (1860) made clear.[14] Indeed, Buckle was greatly influenced by the theological debates centred on the role of Christianity in a rational analysis of humanity. He was firmly on the side of secularists and liberal theologians and he was appalled by the way in which the authors of *Essays and Reviews* were treated as heretics and even contributed to their defence.[15] He was also a firm believer in Charles Lyell's uniformitarian geology against the catastrophic views that garnered favour at Cambridge, though he was disappointed with Lyell's rejection of evolution finding his treatment of the subject 'unsatisfactory'.[16] He was much happier when Darwin's *Origin* was published in 1859, a much more careful and serious treatment of the subject that he completely accepted. Upon reading the book, he wrote to his friend Mrs Woodhead pressing her to tell her 'husband to read *Darwin on Species* and to *master* it. He will find it full of thought and of original matter.'[17]

As influenced as Buckle may have been by the evolutionary naturalists his own scientific history failed to embrace the kind of organic picture of nature presented in the likes of the *Origin*. Buckle's work, rather, was more clearly indebted to the mechanistic views of the French social theorist Auguste Comte (1798–1857). Buckle's debt to Comte was certainly not unique at this time in England. Indeed, many free-thinking Victorian intellectuals from John Stuart Mill and Leslie Stephen to George Eliot, G. H. Lewes and Herbert Spencer, were at one time or another quite excited about some of Comte's ideas.[18] Mill, in particular, before having a general falling out with Comte (as did most Victorian intellectuals over Comte's Religion of Humanity), was absolutely central in popularizing many of Comte's ideas in the 1840s but it was Buckle who actually sought to appropriate and apply Comte's early programme to a given area of study.

Comte, argued Buckle, was not only the greatest writer on the philosophy of method in 'our own time' but he also had 'done more than any other to raise the standard of history' itself.[19] Buckle appreciated that Comte's positivist philosophy was developed out of an understanding of the process of history. Progress has been achieved in the past, argued Comte, by shedding theological ways of thinking in favour of scientific ones. Comte believed that contemporary European society was on the cusp of a 'positivist' stage of history, a stage of ultimate progress whereby knowledge of man would finally benefit from the advancements of science. The creation of a single scientific method that would bring order to the intellectual chaos surrounding the studies of man would go some way toward achieving the positivist goal. In this way, Comte believed, it was possible to uncover laws that could explain society just as astronomers explain the motions of the planets.[20] In the same way, Buckle hoped 'to accomplish for the history of man something equivalent, or at all events, analogous, to what has been effected by other inquirers for the different branches of natural science'.[21]

This meant, for Buckle, realizing that the human past is governed by laws not unlike those found in the natural sciences, that the past is comprehensible by perceiving the 'regular uniformity of sequence' of events.[22] In making history a science, however, Buckle argued that there was a 'preliminary obstacle' that must be overcome and this just happened to be the same obstacle to scientific thought that Comte exposed: theology. According to Buckle, it was very difficult to make history a true science when there is a widespread belief that 'in the affairs of men there is something mysterious and providential, which makes them impervious to our investigations, and which will always hide from us their future course'. Following this logic, or, for Buckle, lack of logic, history could not become a scientific discipline of study. If we consider historical events from a longer time frame, we will understand that historical events are linked inextricably to a series of 'antecedents'. By engaging in such an analysis, the false theories of Providence and free-will will be rejected leading one

> to the conclusion that the actions of men, being determined solely by their antecedents, must have a character of uniformity, that is to say, must under precisely the same circumstances, always issue in precisely the same results. And as all antecedents are either in the mind or out of it, we clearly see that all the variations in the results, in other words, all the changes of which history is full, all the vicissitudes of the human race, their progress or decay, their happiness or their misery, must be the fruits of a double action; an action of external phenomena upon the mind, and another action of the mind upon the phenomena.[23]

For Buckle, then, the progress of civilization was determined by a society's intellect, that is, by a civilization's ability to overcome the forces of nature and shed imaginary belief systems in favour of scientific knowledge. Closely following Comte's stages of human development, the theological, metaphysical and scientific, Buckle argued – with some help from Montesquieu – that in the early stage of human development, the environment dominated the senses. Early societies were simply unable to concern themselves with anything but the twists and turns of the weather and of the desperate need to find food and shelter. Crude theological ways of thinking dominated in such a situation, thereby stalling the progress of knowledge.

Given the poor environment of many parts of the earth, Buckle argued that several civilizations had yet to move beyond this early stage of development, Ireland, Egypt, Central and South America being his foremost examples.[24] The history of Europe excepting Ireland, on the other hand, given its favourable climate, offered an example of how scientific knowledge can blossom where external circumstances are favourable, and Buckle compared various nations of Europe to show where metaphysical modes of thought still held power as well as where scientific knowledge was gaining ground. It was England that provided 'the standard by which we must measure the value of the history of any nation', argued Buckle, who proposed England as his ultimate example of the progress of civilization, a progress that was only being stalled by those who would fail to admit history's scientific status.[25]

Indeed, Buckle himself (again much like Comte) was very much a part of his history, arguing that if in England there was an impediment to finally reaching the scientific/positivist stage it was in the fact that knowledge of humanity was still dominated by theological or metaphysical interpretations and otherwise meaningless history. Since 'there must always be a connexion between the way in which men contemplate the past, and the way in which they contemplate the present', argued Buckle, the status of history was an important marker of a society's progress.[26] However, Buckle found the writing of history to be still 'miserably deficient'.

> Instead of telling us those things which alone have value, – instead of giving us information respecting the progress of knowledge, and the way in which mankind has been affected by the diffusion of that knowledge, – instead of these things, the vast majority of historians fill their works with the most trifling and miserable details: personal anecdotes of kings and courts; interminable relations of what was said by one minister, and what is thought by another; and, what is worse than all, long accounts of campaigns, battles, and sieges, very interesting to those engaged in them, but to us utterly useless, because they neither furnish new truths, nor do they supply the means by which new truths may be discovered.

For Buckle, 'this was the real impediment which now stops our advance'.[27] Buckle's *History of Civilization* was at once an attempt to uncover the laws of civilization's progress but also to contribute to the coming of a truly positivist stage of human development, by making history a science.

While Buckle was clearly influenced by Comte's very general idea of applying the methods of science to that of the study of society – so much so that Buckle is often referred to as a proponent of a 'positivist' science of history – he was by no means simply the 'English Comte' as his opponents often claimed. His first biographer, Alfred Henry Huth, is quite right that many of Buckle's seemingly Comtean moments could have easily been influenced by a whole host of Buckle's favourite French Enlightenment figures from Condorcet on the historical stages of development to Montesquieu on the role of the environment in the historical process.[28] Even his rejection of metaphysical explanations in historical analysis could just as easily have been influenced by the growing literature on biblical criticism and broader theological debates that had embroiled the Church of England throughout the 1840s. What is more, much like Mill and other supporters of Comte's early work, Buckle was appalled by the shift in Comte's thinking from a strictly scientific analysis of society to what was called a Religion of Humanity that sought to order society along scientific principles while embracing many of the conventions associated with organized religion. Buckle believed that Comte may have been 'the most comprehensive thinker France has produced since

Descartes' and yet he sought to organize 'a scheme of polity so monstrously and obviously impracticable, that if it were translated into English, the plain men of our island would lift their eyes in astonishment, and would most likely suggest that the author should for his own sake be immediately confined'.[29] As radical as Buckle was in certain respects, when it came to changing social structures he was much more willing to let nature take its course, seeking to uncover the laws that could help society understand that course rather than alter it for some grandiose utopian vision. It is, therefore, somewhat anachronistic to refer to Buckle's programme as 'positivist' as most Victorians would have associated that term with Comte's Religion of Humanity as well as his programme of applying the methods of natural science to that of society. And while Buckle may have followed Comte in the very general sense of wanting to appropriate scientific methods for the analysis of history, his own science of history was not confined to Comtean principles and sources in establishing his own fairly heterodox scientific method.

For example, in order to uncover history's regularities, Buckle believed that such was made possible by recent gains in the science of statistics. Buckle was a great believer in the statistical method of Belgian Adolphe Quetelet (1796–1874), who was, in Ian Hacking's words, 'the greatest regularity salesman of the nineteenth century'.[30] In examining statistical data concerning social phenomena, Quetelet argued that there tended to be a statistical stability over the long run despite variations over the short term. Buckle believed, with the help of recent statistical studies, the true uniform nature of the past could be uncovered.[31]

Much of the first chapter in the *History of Civilization* is taken up with illustrations of recent findings in the science of statistics, in order to point out the determining influence of hidden laws on the action of people. Not only are a consistent number of people going to kill themselves on a regular basis, but, as Buckle explained, the regularity of this exact number over many years suggests that there is nothing anyone can do to stop it. 'In a given state of society', argued Buckle, 'a certain number of persons must put an end to their own life. This is the general law; and the special question as to who shall commit the crime depends of course upon special laws; which, however, in their total action, must obey the large social law to which they are all subordinate.' Buckle went on to emphasize that because 'the power of the larger law is so irresistible, that neither the love of life nor the fear of another world can avail anything towards even checking its operation'.[32] Buckle also provided similar evidence to show the regularity of crime in France, the regularity of marriage rates depending on the price of corn in England, and the uniformity of letter-writers, who, year after year, through forgetfulness, failed to include an address. Given such evidence, argued Buckle, 'there must be an intimate connexion between human actions and physical laws'.[33]

Buckle believed that underneath the surface differences found in the past one could uncover a deeper uniformity governing and regulating human actions. As Buckle believed such actions were determined by the intellect, he argued that 'intellectual conduct is regulated by the moral and intellectual notions prevalent in their own time'.[34] In this regard, Buckle was clearly influenced by Quetelet's 'average man', a man who represents the mean of a group. Quetelet largely referred to *un homme type* in regard to external qualities such as height, but Buckle believed that the intellect of a society could be understood in much the same way. It was possible for individuals to 'rise above' the intellectual notions prevalent in their own time in the same way that it was possible for 'many others to sink below them'. Such cases, argued Buckle, are the exceptions that prove the rule. The 'immense majority of men must always remain in the middle state, neither very foolish nor very able, neither very virtuous nor very vicious, but slumbering on in a peaceful and decent mediocrity, adopting without much difficulty the current opinions of the day, making no inquiry, exciting no scandal, causing no wonder, just holding themselves on a level with their generation, and noiselessly conforming to the standard of morals and of knowledge common to the age and country in which they live'. In this way, Buckle was not concerned with the action of great individuals in his study but the 'great average of mankind'.[35] As he explained to a correspondent who criticized his failure to treat individuals, 'my inquiry has nothing to do with the individual, but is solely concerned with the masses'.[36]

It was Buckle's matter-of-fact denial of free will and the power of laws to determine the actions of individuals that got people reading the book beyond the usual group of upper-class readers. And people were not disappointed when they picked up the book. 'Enthusiastic young ladies went about "panting for wider generalizations"', argued Leslie Stephen, 'and the general reader was agreeably thrilled by the statement that a mysterious fate might at any moment force him to commit a murder in order to make up the tale required by the law of statistics'.[37] While the general reader might have been amused by Buckle's 'science of history', English historians were decidedly not. As we shall see in the next chapter, they were not happy about his derogatory smears about the trifling state of historical knowledge nor were they impressed by his rejection of – in the same breath – both free will and Providence. It was easy to caricature Buckle's work, and historians proved quite capable of making Buckle look like a radical positivist blinded by his master's 'infidel philosophy', to use Lord Acton's terms.[38]

Any thorough reader of Buckle, however, knows that the book does not follow the rather mechanical method set out in its opening pages. While Buckle claims in those early pages that it is statistics and laws that are the scientific historian's new bread and butter, it is the familiar terrain of individuals and their struggles that follow, though Buckle's analysis largely concentrates on the realm of ideas rather than actions or events.

While his broad arguments usually consider society at large, or what Buckle calls the masses, his focus on progress and inevitable historical change creates a tension between Buckle's supposed concentration on the statistical mean of a society and the impetus for change that must then originate from within what appears to be a changeless uniformity. To this line of questioning Buckle's answer would be found wanting: that there is a constant 'spirit' of scepticism within societies that have overcome nature's external circumstances. It is not, however, from within Buckle's all important masses where intellectual change is engendered, but only by the impetus of 'great men' who exist as if outside of history. While Buckle claims only to be concerned with the masses he takes a Carlylean turn towards great individuals in order to explain how society progresses. Great men

> are immortal, they contain those eternal truths which survive the shock of empires, outlive the struggle of rival creeds, and witness the decay of successive religions. All these have their different measures and their different standards; one set of opinions for one age, another set for another. They pass away like a dream; they are as the fabric of a vision, which leaves not a rack behind. The discoveries of genius alone remain: it is to them we owe all that we now have, they are for all ages and all times; never young, and never old, they bear the seeds of their own life; they flow on in a perennial and undying stream; they are essentially cumulative, and giving births to the additions which they subsequently receive, they thus influence the most distant posterity, and after the lapse of centuries produce more effect than they were able to do even at the moment of their promulgation.[39]

It is unclear how one squares Buckle's view of great men with his focus on the intellect of the masses as well as his determinism. Buckle claims time and again that men are products of their circumstances but then must turn away from such a claim when trying to explain the changing of those circumstances by employing the concept of genius. His use of statistics to uncover history's average intellect does not mesh with his view of great men who tend to guide the masses in a new direction toward a scientific stage of history.

For Buckle, great men are simply the exception to the general rule that 'the actions of individuals count for very little'.

> Whoever is accustomed to generalize, smiles within himself when he hears that Luther brought about the Reformation; that Bacon overthrew the ancient philosophy; that William III. saved our liberties; that Romilly humanized our penal code; that Clarkson and Wilberforce destroyed slavery; and that Grey and Brougham gave us Reform. He smiles at such assertions, because he knows full well that such men, useful as they were, are only to be regarded as tools by which that work was done which the force and accumulation of preceding circumstances had determined

should be done. They were good instruments; sharp and serviceable instruments, but nothing more.

This was why, for Buckle, it was so important to view history from the perspective of a long period, while searching for the deeper similarities hidden underneath the surface differences. In this way great men will appear, not on a regular basis, but their legacies extend beyond centuries, as if outside of history. Under such a conception, however, '[t]he occurrences which contemporaries think to be of the greatest importance, and which in point of fact for a short time are so, invariably turn out in the long run to be the least important of all. They are like meteors which dazzle the vulgar by their brilliancy, and then pass away, leaving no mark behind.'[40]

It was a difficult task for historians to see the truly remarkable and important facts, argued Buckle, and he proposed a rather peculiar method for seeing beyond the surface differences in order to uncover those truly great moments that not only dazzled contemporaries but also shaped the historical process for centuries to come. It was for largely this reason that Buckle abhorred what he called the 'inductive tendencies' of the nineteenth-century English mind which seemed to think that only a minute study of particular facts could produce scientific knowledge of the world. This way of thinking for Buckle applied to both men of science and historians who for Buckle seemed to embrace Baconianism with an 'almost superstitious reverence'.[41] Here Buckle was echoing a growing strain of criticism that was being voiced against a rather simplistic Baconianism, an argument that also finds resonance in the very different work of Romantic historian T. B. Macaulay and physicist David Brewster.[42] However it was John Stuart Mill who perhaps influenced Buckle more than anyone else when it came to debates about the inductive method in mid-Victorian Britain.

Buckle was a very close reader of Mill's *System of Logic* (1843) and he seems to have agreed with Mill when it came to the latter's interpretation of the inductive method, that it was a method that concerned the strict observation of facts devoid of any instinctual *a priori* connotations. Buckle clearly believed that the two men had a very similar view of both science and society as he was convinced Mill's forthcoming *On Liberty* (1859) would accord with his own views on the subject, so much so that he approached Parker about reviewing the book despite his already heavy work load that was clearly worsening his illness.[43] *On Liberty*, it would turn out, would not disappoint Buckle, and he reviewed it as promised, though his health was at this point in rapid decline. The influence was by no means mutual, though Mill certainly enjoyed Buckle's work and he even added a fairly positive section on Buckle's *History* in later editions of his *System of Logic*, though he was not uncritical.[44] He had even apparently considered editing what was left of Buckle's *History*, that perhaps a posthumous volume could be put together from Buckle's well-kept notes. This was not to be, however, and Buckle's friends and dwindling fans would have to settle for reading his common place books and other miscellaneous writings in print nicely edited by Mill's daughter, Helen Taylor.[45] For Mill, Buckle 'was performing a most valuable function in popularising many important ideas, and stimulating the desire to apply general principles to the explanation and prediction of facts', but he was suspicious of the 'undue breadth of many of his conclusions, and the want of a proper balance in his mind'.[46] According to Mill, Buckle was more a popularizer of his own fashionable ideas, rather than an originator of them, and where Buckle's thought moved away from Mill's own, he was more likely to disagree than agree with him.

Buckle was no blind follower of Mill either. When it came to politics, he was much more inclined to agree with classical economic theorists such as Adam Smith and Thomas Malthus than he was with Mill. As we have seen with his criticisms of Comte's Religion of Humanity, he wanted to let history take its course, rather than force supposed liberal reforms on an unwilling public. He was very suspicious of that utilitarian streak in Mill's work, that 'old doctrinaire school' that Mill 'never quite got rid of'. 'The traditions of that school were handed down to him by his father direct from Jeremy Bentham; and tho' Bentham was one of the most eminent thinkers this or any other country has ever possessed, he was so unversed in the *art* of life (as distinguished from the *science*) that if he had possessed the requisite power he would have inflicted more misery upon England than has ever been inflicted on it by any single man – meddle, meddle, meddle, is always the cry of the speculator.' Buckle, sounding very much like a crusty Edmund Burke, argued that often perceived ills lead to needless reforms that would inevitably provoke reactions far more harmful than the original problem. 'The history of human affairs, in modern times, is the history of these reactions, all of which have been full of danger – and none of which would have occurred, if men would bide their time, and would only condescend to *sap* bad institutions before they try to overthrow them.'[47]

As it is well known, however, Mill's own thinking took a Wordsworthian turn when he began to read Romantic poetry and literature in the wake of a nervous breakdown brought on by his father's utilitarian education regimen as prescribed by Bentham. Bentham may not have been versed in the 'art of life' as Buckle put it, but Mill eventually broke free from his utilitarian clutches and found solace in the sublime poetry of William Wordsworth in particular. It was necessary, according to Mill, to find a balance between what he saw as a great divide in contemporary English thought between the mechanistic utilitarian science of Bentham on the one hand and the Romantic sensibilities of Samuel Taylor Coleridge on the other.[48]

For Buckle, however, Mill did not go quite far enough in this regard. While Mill embraced the need to incorporate a poetic sensibility with a rational one, he was still for the most part on Bentham's side of the cultural divide and such is particularly apparent in his rather narrow view of the inductive method. It just so happens that a more Coleridgean view of the inductive method was promoted by the premier theorist of induction, William Whewell. *Pace* Mill, Whewell argued that induction had historically been practiced not just by observing facts that would lead to larger generalizations but by also relying on barely understood traits of the human mind, particularly that of intuition, that would allow the scientific observer to make accurate speculations that went beyond the strictly observed phenomena. Mill was opposed to this kind of thinking, not just because of an underlying utilitarian streak, as Buckle might argue, but because he saw the argument in favour of an inductive intuition in particular as leaving open a space for the supernatural or divine guidance in human understandings of nature.[49] Buckle was less concerned with the political implications of believing in some form of intuition in interpreting both nature and the past. In order to get beyond the individual facts that might present a very short-sighted view about what was truly important about a given age, the scientific historian for Buckle would need to rely on a whole host of intellectual operations from a form of strict observation to an embrace of poetry and the imagination, to speculation and intuition.

Interestingly, Buckle seems to have agreed with Mill's rather narrow and mechanistic definition of the inductive method and perhaps that is why he found it so wanting as a stand-alone method in the analysis of history. Instead of turning towards Whewell's more nuanced view of induction that would make a space for such Coleridgean traits as the imagination and intuition, Buckle argued that induction, on its own, was simply insufficient and not just for the science of history but for science itself. He was certainly aware of Whewell's work and his common place books and notes for his *History* show that he relied quite heavily on Whewell's *History of the Inductive Sciences* (1837) and his *Philosophy of the Inductive Sciences* (1840).[50] But Buckle seems to have been prejudiced against Whewell because of the latter's 1833 Bridgewater Treatise where it is argued that induction will not only lead to truth but to the ultimate Divine Truth about the Supreme Deity. There are a few condescending remarks about that work that appear in Buckle's *History*.[51] And yet Buckle's own interpretation of some of the great scientific discoveries of the English past seems to owe more to Whewell than to Mill.

In the second volume of the *History of Civilization in England* (1861), there is a lengthy section on the development of modern science where Buckle argues for the centrality of the imagination and in particular of poetry in the discovery of central scientific laws. 'There is, in poetry', argued Buckle, 'a divine and prophetic power, and an insight into the turn and aspect of things, which, if properly used, would make it the ally of science instead of the enemy'. The poet contemplates nature by relying on his 'emotions'; the man of science contemplates nature through 'understanding'. 'But the emotions', Buckle explained, 'are as much a part of us as the understanding; they are as truthful; they are as likely to be right'. Poetry, Buckle went on to explain, 'is a part of philosophy, simply because the emotions are a part of the mind. If the man of science despises their teaching, so much the worse for him. He has only half his weapons; his arsenal is unfilled.'[52]

In his analysis of the English scientific discoveries of the seventeenth century, Buckle sought to make explicit links between both the science and the poetry of the time. He claimed that he could 'hardly doubt' that the 'wonderful' scientific discoveries of Newton and Harvey were made 'because that century was also the great age of English poetry'. The great poets helped create a cultural milieu that would allow the English mind to transcend its previous limitations.

> The two mightiest intellects our country has produced are Shakespeare and Newton; and that Shakespeare should have preceded Newton was, I believe, no casual or unmeaning event. Shakespeare and the poets sowed the seed, which Newton and the philosophers reaped. Discarding the old scholastic and theological pursuits, they drew attention to nature, and thus became the real founders of all natural science. They did more than this. They first impregnated the mind of England with bold and lofty conceptions. They taught the men of their generation to crave the unseen. They taught them to pine for the ideal, and rise above the visible world of sense.

Buckle was essentially arguing that without poetry, science would not have progressed to the extent that it has. '[B]y cultivating the emotions', argued Buckle, poetry 'opened one of the paths to truth', a path that would have remained otherwise closed.[53]

These rather tenuous links Buckle sought to maintain between a poetic imagination and scientific progress are somewhat clearer in his famous public lecture of 1858. The lecture that had 'everybody in London talking' was on 'The Influence of Women on the Progress of Knowledge', a seemingly odd title for a talk that was supposed to be about the science of history. But it was less about women's influence on knowledge than on the genders of scientific knowledge and the talk itself is a wonderful illustration and comment on the way in which gender informed conceptions of knowledge at the time.[54]

Buckle began his talk by admitting that women's direct influence on the historical process has been quite limited, particularly if the focus is on the progress of knowledge. '[T]o state the matter candidly, it must be confessed that none of the greatest works which instruct and delight mankind have been composed by women. In poetry, in painting, in sculpture, in music, the most exquisite productions are the work of men.' If we consider women's influence on the progress of knowledge solely on the merits of great works, we should infer, as has been 'openly stated by eminent writers, that women have no concern with the highest forms of knowledge; that such matters are altogether out of their reach; that they should confine themselves to practical, moral and domestic life, which it is their province to exalt and beautify'. From such an argument, it is often concluded that women should be restricted to the private sphere, that 'field of their really useful and legitimate activity'.[55] This argument, for Buckle, was a straw man that he set up to take down dramatically.

Against this received wisdom, Buckle claimed that women have and do exert a profound influence on the progress of knowledge by keeping alive a way of thinking that men have largely abandoned. '[T]he point I shall attempt to prove', argued Buckle, 'is that there is a natural, a leading, and probably an indestructible element, in the minds of women, which enables them, not indeed to make scientific discoveries, but to exercise the most momentous and salutary influence over the method by which discoveries are made'.[56] In this way, women have contributed to the greatest scientific discoveries in history by keeping alive a method of analysis that is better suited to woman's nature than it is to man's. Buckle was referring to the method of deduction.

Buckle defined the deductive method, rather simplistically, as the method that begins not with facts but ideas. This was in contradiction to the more masculine method of induction, which begins not with ideas but with facts. 'The inductive philosopher collects phenomena either by observation or by experiment, and from them rises to the general principle or law which explains and covers them. The deductive philosopher draws the principle from ideas already existing in his mind, and explains the phenomena by descending on them, instead of rising from them.'[57] Because of their nature, women are more suited to the deductive method, while men are more suited to the inductive method. Both methods, however, are necessary to employ in any scientific investigation. In other words, the most scientific of knowledge can only be produced by combining the inductive and deductive methods in any given analysis.

If it were not for women, Buckle believed, the inductive method would have long triumphed as the only method of scientific analysis, which is to say that there would no longer be any true scientific method. Women have encouraged men to maintain the 'deductive habits of thought' and have therefore prevented 'scientific investigators from being as exclusively inductive as they would otherwise be.' This, therefore, has been women's great 'though unconscious, service to the progress of knowledge'. That women had yet to produce a truly great work of art or scientific discovery says nothing concerning their importance in the progress of knowledge. Indeed, Buckle argued, women's preference for deduction kept true scientific analysis and progress alive.[58]

In explaining the difference between men's and women's patterns of thinking, Buckle trotted out all of the familiar tropes concerning gender found in Victorian society. Women prefer the deductive method because they 'are more emotional, more enthusiastic, and more imaginative than men; they therefore live more in an ideal world'. Men, on the other hand, 'are more practical and more under the dominion of facts'. Women also, and in contrast to men, 'possess more of what is called intuition. They cannot see so far as men can, but what they do see they see quicker.' For Buckle, this suggested that women are able to 'grasp at once at an idea, and seek to solve the problem suddenly, in contradistinction to the slower and more laborious ascent of the inductive investigator'.[59] While this train of essentialist thinking tended to justify contemporary gender stereotypes – that women are naturally not scientific thinkers – while simply turning the usual inference from it on its head, Buckle went on to give evidence of this type of feminine thinking at work in the past in the most important of scientific discoveries. Buckle believed that it was, in fact, deductive thinking that led Newton to the discovery of gravity.

Newton had to think in both masculine (inductive) and feminine (deductive) ways to produce the scientific laws that shaped so much in the Victorians' world. Buckle went through the famous apple falling out of the tree routine to help explain how Newton produced the law of gravity.

> Observe how he went to work. He sat still where he was, and he thought. He did not get up to make experiments concerning gravitation, nor did he go home to consult observations which others had made, or to collate tables of observations: he did not even continue to watch the external world, but he sat, like a man entranced and enraptured, feeding on his own mind, and evolving idea after idea. He thought that if the apple had been on a higher tree, if it had been on the highest known tree, it would have equally fallen. Thus far, there was no reason to think that the power which made the apple fall was susceptible of diminution; and if it were not susceptible of diminution, why should it be susceptible of limit? If it were unlimited and undiminished, it would extend above the earth; it would reach the moon and keep her in her orbit. If the power which made the apple fall was actually able to control the moon, why should it stop there? Why should not the planets also be controlled, and why should not they be forced to run their course by the necessity of gravitating towards the sun, just as the moon gravitated towards the earth?

This wonderful illustration of Newton's thought process was, for Buckle, the most powerful example of the fundamental importance of deductive thinking in the creation of scientific knowledge. Just by sitting under the tree and thinking about what had happened, Newton's mind advanced 'from idea to idea' until

> he was carried by imagination into the realms of space, and still sitting, neither experimenting or observing, but heedless of the operations of nature, he completed the most sublime and majestic speculation that ever entered into the heart of man to conceive. Owing to an inaccurate measurement of the diameter of the earth, the details which verified this stupendous conception were not completed till twenty years later, when Newton, still pursuing the same process, made a deductive application of the laws of Kepler: so that both in the beginning and in the end, the greatest discovery of

the greatest natural philosopher the world has yet seen, was the fruit of the deductive method. See how small a part the senses played in that discovery! It was the triumph of the idea!⁶⁰

Had Newton merely proceeded by induction, he would have never moved beyond the apple and the tree. It was to the benefit of knowledge and science that Newton was moved by a deductive spirit, argued Buckle. If it were not for women 'scientific men would be much too inductive, and the progress of knowledge would be hindered'. The law of gravity is just one example of women's great 'unconscious service to science'.⁶¹

This is not the place for an analysis of the gendered conceptions of knowledge during the Victorian period, though Buckle's lecture does pose a problem to the general thesis that making history a science necessarily meant making history masculine.⁶² What is worth mentioning, at this stage, is that Buckle's conception of science itself is a heterodox one, an odd mixture of a rather mechanistic conception of induction and the past combined with an embrace of intuitive and imaginative sensibilities. Buckle wanted the scientific historian to discover laws that govern the human past, but gaining that knowledge for Buckle could not be achieved by merely observing a mountain of facts. The true scientific historian had to rely on the imagination, on poetry, on emotions, as well as on observed facts, in order to establish the laws of history. We need to 'give more scope to the imagination', argued Buckle, 'and incorporate the spirit of poetry with the spirit of science'.⁶³

Buckle's method was just a little too heterodox for some people. While he and Whewell seem to agree on a very general level about how scientific discoveries are made, Whewell found Buckle's definitions of induction and deduction made only to suit the latter's particular arguments about women rather than about science.

> He opposes to Induction, which he says is the male habit of mind, what he calls Deduction, which he says is a better thing which women have. But by Deduction he means Induction, and such Induction as is a necessary part of all Inductive discoveries. And so he practices the common trick of changing the meaning of words, and then startling you by a paradoxical assertion. So you see I am not going to admire women for *his* reasons, thinking that I have better of my own for so doing.⁶⁴

For Whewell, Buckle's method that combined induction and deduction was simply his own inductive approach under a new name. Charles Darwin, who was greatly influenced by Whewell's philosophy of method as is perhaps best exemplified by first page of the *Origin* where Whewell is quoted in an epigraph,⁶⁵ found himself and his friend and great collaborator Joseph Hooker arguing with Buckle at a dinner party over precisely his faulty definitions of induction and deduction. Darwin thanked Hooker for the way in which he 'stuck up about induction and deduction'. Hooker, meanwhile, was just happy when the conversation came to an end: 'As to my argument with him I felt doubled up altogether – the cool way in which he 3 times shifted his ground gave me the worst opinion of his unfairness & when he said "Ah – well – perhaps we have different ideas of what the Inductive & Deductive methods are" – I felt I had got my belly-full & sneaked away as I should from a treacherous savage, looking over my shoulder to see if he followed.'[66]

In hindsight Buckle's was an odd but popular message. It took hold of the popular imagination in a way that would have seemed impossible a few years before. Buckle certainly took advantage of a milieu conducive to scientific historical thinking but there was something about the way in which his book was written that made it accessible to different classes of readers. In this way there are interesting parallels with Robert Chambers's anonymously published *Vestiges of the Natural History of Creation* (1844). Like Buckle's *History*, Chambers's *Vestiges* took advantage of a particular milieu, his just happened to be one where evolutionary theory was ripe for consideration, but the book was written in a wonderfully Romantic style that owed quite a bit to the great Romantic historical novelists of the first half of the nineteenth century. It was a book that engendered a whole host of readings depending on the particular reader, and yet there was a fair amount of consensus from within the scientific community that the book was at best bad science and at worst simply heretical.[67] Buckle's book invoked similar responses and was also quite popular in part because of its style.[68]

Indeed, Buckle's prose was an absolute pleasure to read. Even though Darwin disagreed with Buckle's interpretation of the scientific method, he still enjoyed reading the *History of Civilization* and thought Buckle to be 'the very best writer of the English language who ever lived'.[69] 'We have writers whose sentences linger in the mind', argued the secularist leader, George Holyoake, 'but we have none who have Buckle's passionate eloquence'.[70] Buckle famously spent much time constructing his sentences searching to find just the right turn of phrase to put his point across as best as he could. He felt that particularly in England, scholars used horrendous language to convey their ideas, and, sounding very much like present-day critics of poststructuralism, argued that they have 'corrupted the English language with a jargon so uncouth, that a plain man can hardly discern the real lack of ideas which their barbarous and mottled dialect strives to hide'.[71] Buckle sought to avoid writing in a difficult style in part because he wanted his book to be popular and read by the lower and working classes. 'I want my book to get among the mechanics' institutes and the *people*; and to tell you the honest truth', he wrote to his friend Emily Shirreff, 'I would rather be praised in popular and ... *vulgar* papers, than in *scholarly* publications'. He went on to suggest that 'vulgar' publications cannot judge the 'critical value' of his work, but they are, more importantly, 'admirable judges of its *social* consequences among their own class of readers. And these are whom I am now beginning to touch, and who I wish to move.'[72]

From the very beginning of conceiving of his *History*, it was important for Buckle to get his message out to the widest possible audience. When he approached Parker about publishing the manuscript, this was his primary concern in making any revisions. He suggested that Parker ask John Foster or Baden Powell to read the manuscript not so much to judge the book's evidence of learning but rather to suggest how the book could be revised in order to make it more popular. Foster was suggested in particular because 'as editor of the *Examiner*', Buckle explained, he 'has, of course, peculiar facilities for judging if a book is likely to be popular'. Buckle doubted that Parker should concern himself about the knowledge displayed in the book given that he had 'been engaged incessantly on it for fourteen years'. The main question to ask any readers would be, rather, 'have I written the book clearly and popularly?'[73] He later wrote to Parker re-emphasizing the importance of considering the book's likely popularity in evaluating its worth:

> the point on which alone you will require information is as to the clearness and attractiveness of the style, which, as a matter of business will be your principal consideration. For if the style is judged to be good, as well as the facts curious, a tolerable success is certain: since every book which has failed has owed its failure either to want of industry in collecting evidence, or else to want of lucidity in arranging it.[74]

Buckle wanted Parker to understand that he 'was deeply impressed with the importance of a clear and popular style and that I have made great and constant efforts to attain it'.[75]

Underpinning much of Buckle's theory was the necessity of having a scientific understanding of society disseminated throughout the population, that such would enable true progress to take place. But there was obviously something quite conceited about Buckle's desire to be so popular that went beyond the justifications of his proposed science of history. Indeed, Buckle wanted to be like one of the few great individuals that appear so rarely from within the static mass of civilization in order to help shape and direct future progress, not through prescriptive social reforms as was Comte's strategy, but rather by popularising a scientific view of the human past that would simply increase society's self-knowledge. Buckle's desire to break free from the deterministic laws of history and become one of the few great men is particularly apparent, ironically, just as he was becoming conscious of his own mortality as his health began to fail and the conclusion of his project began to fade from his view. In 1856, when his health took another downturn, he told Maria Grey that he was terrified that he might 'break down in the midst of what according to my measure of greatness is a great career ... and the thought of which seems to chill my life as it creeps over me'. The thought entered his mind that perhaps he had 'aspired too high' but at this point he still had 'such a sense of power, such a feeling of reach and grasp, ... such a command of the realm of thought, that it was no idle vanity to believe that I could do more than I shall now ever be able to effect'.[76]

By the time of writing his second volume, however, this momentary lapse in self-confidence became all-embracing. He wrote to Harriet Grote that he was 'miserably nervous and tormented by the thoughts of how little I can do, and how vast an interval there is between my schemes and my powers'.[77] This message is all too clear in the second volume itself where he admits that his project will be impossible to complete during his lifetime. 'Once, I own, I thought otherwise. Once, when I first caught sight of the whole field of knowledge, and seemed, however dimply, to discern its various parts and the relation they bore to each other, I was so entranced with its surpassing beauty, that the judgement was beguiling, and I deemed myself able, not only to cover the surface, but also to master the details.' But Buckle had to admit that he was wrong. 'Little did I know that the horizon enlarges as well as recedes, and how vainly we grasp at the fleeting forms, which melt away and elude us in the distance.'[78] This is not the language of a man deluded by the positivist dream of absolute knowledge.

Buckle would spend the last year of his life travelling throughout the Middle East seeking a warmer climate to help treat his rapidly worsening illness. For much of his trip he seemed to be recovering quite well and was rejuvenated to get back at work on his *History*. 'I must tell you', he wrote to Mrs Huth on 16 April 1862, 'that I am far stronger both in mind and body than I have been since you knew me, and feel fit to go on at once with my work'.[79] Just six weeks later, however, on 29 May 1862, Buckle would be dead. He contracted typhoid fever in Beirut on his way to Damascus but refused to halt his journey. In the words of the acting Consul in Damascus, when Buckle was bed-ridden, the French doctor that was called to attend him employed the 'lowering system of treating fevers in vogue with his countrymen, and had already bled and leached his patient freely, allowing him only cooling drinks', a treatment that surely worsened Buckle's weakened state. The Consul telegraphed Beirut for a more competent American doctor who 'at once administered stimulants and nourishing broth, but the patient never rallied, nor did consciousness return'.[80] He would soon after be buried in a Damascus cemetery and only the sustained efforts of a few acquaintances assured that a proper tomb would mark his final resting place.

After an initially welcomed reception, particularly among the group of scholars associated with the freethinking *Westminster Review*, George Henry Lewes, John Stuart Mill, George Eliot, Leslie Stephen, and editor John Chapman, the same group initially attracted to Comte's work in the 1840s and 1850s, Buckle's project began to be pilloried in the periodical press, most notably by historians who used his work as a foil to promote their own brand of scientific history or to reject the science of history *in toto*.[81] Even his early proponents began to agree with the critics and were certainly unwilling to defend his views after his death. George Eliot made that particularly clear in a scathing 1865 review of the only work of history that was truly indebted to Buckle.[82] From this perspective, Buckle has more in common with the much-maligned theological historian Robert William Mackay (1803–82) than with the anonymous author of *Vestiges*. Mackay, much like Buckle, was a wealthy independent amateur who made an initial splash with a book embracing fashionable views of biblical criticism. He was, just like Buckle, initially embraced by the *Westminster* crowd highlighted by an enthusiastic review from George Eliot in 1850,[83] only to be soon after rejected as at best a mere popularizer in the very same journal that had at once embraced him as an original thinker.[84] As a result Mackay's work was soon forgotten. The difference would be that while Buckle's work was also rarely taken seriously beyond the early 1860s, it would still 'remain long after the writer', according to the Arabic inscription on Buckle's renovated tomb stone,[85] if only because it provided the occasion for a formative debate about history's scientific and professional status. As late as 1873, the *Westminster Review* admitted that Buckle's surprisingly popular first volume 'Exercised an incalculable influence upon the whole science of history. No views of history can ever be held again without being affected in one way or another by that commanding work.'[86] Indeed, despite the fact that Buckle did not accomplished his immense task of making history a natural science, of essentially changing the course of history like the few unique individuals that appear in his *History of Civilization*, it is safe to say that while he certainly dazzled his contemporaries for a few short years, he did not fail to leave a mark behind.

2 THE SCIENCES OF HISTORY

Method, not genius, or eloquence, or erudition, makes the historian.

Lord Acton, 'Mr. Goldwin Smith's *Irish History*', *Rambler*, 6 (1862)

Even though the early reaction to Buckle's work was generally mixed, the burgeoning historical community was much more willing and able to reach a consensus opinion. 'Have you seen Buckle on Civilization Vol. 1[?]', historian William Stubbs (1825–1901) wrote to his friend and fellow historian, Edward A. Freeman (1823–92). 'There are to be Ten. I do not believe in the Philosophy of History so do not believe in Buckle'. He went on to imply, however, that he assumed Freeman would like the work as it promoted history to the ranks of science, a proposition that Stubbs himself found attractive. He simply found Buckle's science of history to be more philosophy than it was science. 'I fear you will make me out to be a heretic indeed after such a confession.'[1]

However, it would be Buckle's science of history that would be denounced as heretical while Stubbs's opinion of Buckle's work would become the orthodox response from English historians. At issue was not whether or not history should become a science. This was a moot question for most. At issue, rather, was what *kind* of science history should become, what *kind* of methods history should employ, and what *kind* of identity the historian should adopt in order to be trusted to impart historical and scientific knowledge to peers and to the wider public. Indeed, Buckle's work engendered much discussion about the kind of science historians' wished to adopt in making their discipline one among the sciences. However, it is clear that Buckle's science of history was not the kind most other historians such as Stubbs had in mind.

At a time when state archives were opening throughout Europe and historians were gaining unprecedented access to the documentary traces of the actual past,[2] Buckle seemed to suggest that the facts themselves were unimportant and that, perhaps, there were simply too many of them.

We are in that predicament, that our facts have outstripped our knowledge, and are encumbering its march. The publications of our scientific institutions, and of our scientific authors, overflow with minute and countless details, which perplex the judgment, and which no memory can retain. In vain do we demand that they should be generalized, and reduced to order. Instead of that, the heap continues to swell. We want ideas, and we get more facts. We hear constantly of what nature is doing, but we rarely hear of what man is thinking. Owing to the indefatigable industry of this and the preceding century, we are in possession of a huge and incoherent mass of observations, which have been stored up with great care, but which, until they are connected by some presiding idea, will be utterly useless. The most effective way of turning them to account, would be to give more scope to the imagination, and incorporate the spirit of poetry with the spirit of science.[3]

For someone like Stubbs, this was absurd. History had to be built on a 'painstaking' foundation of minute facts and the more facts the better. As he explained to Freeman in a 'Baconian Simile' (that could have been written by Dickens in the voice of Thomas Gradgrind) in describing the method that underpinned his newly published study of Episcopal succession, *Registrum sacrum Anglicanum* (1858): 'All chronological minutia are the pebbles of the concrete in which the foundations of history must be laid'. Stubbs was 'particularly anxious to have understood ... that it is a work of original research not a réchauffé' of work previously done by other scholars.[4] Here Stubbs was promoting a very different science of history than that offered by the author of the *History of Civilization in England*, and one that had the benefit of being derived from a seemingly English tradition stretching back to Francis Bacon as opposed to the recent French one Buckle appeared to appropriate. Any new science in nineteenth-century Britain, whether of the physical science variety such as geology or a more human science such as history, had to be, in the words of Antonio Pérez-Ramos, 'ceremonially Baconian if it aspired to respectability'.[5] Buckle's was ceremonially *anti*-Baconian.

Many historians of Victorian history-writing have failed to notice the distinction between Buckle's science of history and that offered by the English historians who would give shape to the nascent discipline of history in Britain, historians such as Stubbs. Because of this, there is a misconception that English historians of the second half of the nineteenth century were thoroughly trained under the guidance of the positivist method.[6] Despite the intervention by Christopher Parker, historians have continued to confuse 'positivism' with 'empiricism' and even 'science' itself.[7] For instance, after suggesting that Buckle advanced the 'idea that history could be a science', Michael Carignan goes on to say that Stubbs's work 'established the standards of positive science by which the new profession policed itself' and that the 'move toward professionalization of history ... followed the trajectory established by Buckle and Stubbs'.[8] Carignan is correct to state that Stubbs's work established an ideal standard for the discipline of history, but it was no 'positive' science. It was, in fact, developed in direct opposition to the type of positivist science of history offered by Buckle. T. W. Heyck put it best when he said that Buckle taught English historians 'to see what they did *not* mean by science', that he provided a negative example of the science of history.[9]

In England, John Emerich Edward Dalberg-Acton (1834–1902), later and better known as Lord Acton, was the first historian to attempt to articulate an alternative to Buckle's science of history in a pair of reviews that appeared in the *Rambler* in 1858, a Catholic-oriented journal that was co-edited by Acton and Richard Simpson. It was Acton's purpose to use Buckle's *History of Civilization* as a foil to put forward his own method of history, one that would embrace the spirit of science while also making a space for Christian metanarratives on one hand and free will on the other. Acton's reviews should be seen as an act of 'boundary work', a concept Thomas Gieryn employs in a similar context to describe 'an ideological style found in scientists' attempts to create a public image for science by contrasting it favourably to non-scientific intellectual or technical activities'.[10] Here boundary work is understood as an attempt to create a public image for history by contrasting it favourably to other supposed historical activities deemed non-scientific.

Acton came from a long-line of European aristocrats. He was born in Naples, the son of Sir Richard Acton and Marie Dalberg who was the daughter of the Duke of Dalberg, the product of a line of South German Catholics living in France. Sir Richard suddenly died of pneumonia on 31 January 1837, leaving the three-year-old Acton without a father and Marie without a husband. This situation did not last long as Marie married George Leveson-Gower, the future Earl of Granville (1846), in July of 1840. This meant that Acton grew up in a cosmopolitan environment speaking French and German despite spending most of his childhood in London. He was also pushed and pulled by various familial influences. The Actons, Dalbergs and Levesons all had different educational plans for the young Acton. He was initially sent to Oscott College until he was sixteen. He then applied to three colleges at Cambridge, but was turned down by all three, thanks to the exclusionary laws regarding non-Anglicans that would not be changed for another few decades. After a variety of possible scenarios were proposed by members of his family, it was agreed that Acton would complete his education in Germany where he lived and worked with a familial acquaintance, Johann Joseph Ignaz von Döllinger (1799–1890) of Munich University, studying ecclesiastical history and theology.[11]

The discipline of history was in a much more advanced state in Germany than in England at the time. Leopold von Ranke was in the midst of becoming the spokesman for a new method of history, one that taught the historian not to judge past actions or instruct 'the present for the benefit of future ages', as had been the typical practice of historical writers, but 'only to show what actually happened' (*wie es eigentlich gewesen*).¹² Ranke spoke of the importance of critical research, of the primacy of facts, of beginning with the particular before moving on to the general and, most importantly, of the necessary 'impartiality' (*unparteylich*) of the historian.¹³ While Ranke was certainly not expounding any new principles to the study of history, he, as Leonard Krieger puts it, 'converted the principles which had been the tenets of individual historians into a paradigm which could be communicated to an entire profession as its distinctive collective identification'.¹⁴ Georg G. Iggers would want to remind us, however, that often ignored in Ranke's paradigm are his writings that suggest more nuance in his method, that he in fact believed (sounding very much like Buckle) that history was both a science and an art, that the historian had to rely on a sort of intuition (*Ahndung*) to make sense of the facts, that he must be guided, in other words, by some overarching transcendent idea.¹⁵ Much like other contemporary English historians, however, Acton seemed to fail to appreciate the nuance in Ranke's work, choosing instead to promote the more obviously Baconian Ranke found in his famous preface to his *Histories of the Latin and Germanic Nations* (1824). Indeed, Acton would thoroughly embrace the somewhat conventional image of Ranke, explaining in his inaugural lecture as Regius Professor of Modern History at Cambridge in 1895 that 'Ranke is the representative of the age which instituted the modern study of History. He taught it to be critical, to be colourless, and to be new. We meet him at every step, and he has done more for us than any other man.'¹⁶

While Acton might have remembered Ranke as his 'master'¹⁷ he was much less impressed with the man after attending one of his famous lectures in 1855 claiming that Ranke was 'very small and ugly, very lively ... and there is no seriousness or dignity in his nature'.¹⁸ Acton, at this early stage in his historical development, was much more influenced by his mentor Döllinger, a Bavarian Catholic whose interpretation of theological history was diametrically opposed to Ranke's clear Protestant leanings. And yet Döllinger and Ranke were guided, at least rhetorically, by a similar method that foregrounded extensive and diligent research while suppressing ideological or religious commitments. Despite what Acton claimed to learn from Ranke in his later years, it is clear that he learned many of the methodological precepts associated with Ranke from Döllinger. What he praised in Ranke, he also praised in Döllinger, for instance, that Döllinger's historical research was guided by 'no purpose' and 'in obedience to no theory, under no attraction but historical research alone' and that Döllinger formed 'his philosophy of history on the largest induction ever available to man'.¹⁹

This is all to say that Acton's formative years were spent in Germany training directly from the accepted masters of historical research. Acton was, because of this, seen as an authority on the German method of historical analysis, a method that was envisioned as providing a foundation for the nascent discipline of history in Britain.[20] The bimonthly *Rambler* provided Acton with a ready-means of communicating that Germanic influence to an English audience.

The *Rambler*'s motto under Acton and Simpson's co-editorship was *Seu vetus est verum diligo, sive novum*, 'I value truth whether it is old or new', a provocative maxim for a Catholic journal that sought to engender discussion about the leading scientific issues of the day. Acton's Catholicism was much like John Henry Newman's (who also worked with Acton on the *Rambler*), in the sense that his faith was not opposed to scientific knowledge. He wanted to convince Catholics that they need not fear scientific advancements and discoveries, that truth, whether pursued by faith-based or secular methods, was the ultimate goal.[21] Foremost on Acton's liberalizing agenda was to introduce the new methods of German scientific historical scholarship to English Catholics.[22] His review of Buckle's work represented a first attempt at doing so.

Given the popular success of Buckle's book, Acton and Simpson believed it was necessary to write a review in *Rambler*, and they initially planned to jointly write a single review. Once Acton read the book, however, he found it to be 'utterly superficial and obsolete' and Buckle himself a 'humbug and bad arguer'.[23] It was decided instead that a pair of reviews would be written in order to use Buckle's 'book as a peg to hang many interesting things upon'.[24] Simpson hoped that it would be possible for Acton 'to expound some of your views of history' contrasting 'Buckle's odd views' with 'your common sense ones'.[25] In the end, it was decided that the *Rambler* would publish consecutive reviews of Buckle's *History of Civilization in England*: the first dealing with Buckle's proposed method of history, with the first draft written by Simpson, though heavily edited by Acton and representing primarily Acton's opinions; the second dealing with Buckle's handling of the empirical evidence, and this was to be more properly written by Acton alone.

The first review on 'Mr. Buckle's Thesis and Method' began by commenting on Buckle's surprising popularity, that he had the 'rare fortune of jumping to celebrity in a single bound', success that is 'far above that which usually attends such efforts'. The book 'must have powerfully appealed to something or other in the public mind, or tell something or other very important, which people wanted to know, in order to have won so rapid a popularity'.[26] That popularity, it is concluded, is due, in large part, to 'the emotions stirred up by Mr. Buckle's eloquence ... [and] grand words' rather than to the substance of those words'.[27] There is, it is argued, an inherent problem in attempting to make the past conform to generalized laws and this is further problematized by logical inconsistencies that plague Buckle's entire programme, a programme that is not really Buckle's, but that of a foreign school of philosophy.

The crux of Buckle's problem, it is argued, is that in order to make the vicissitudes of the past follow some generalized law, Buckle must look at the past as if it is made up, not of individuals, but of large bodies, e.g., groups, classes, mobs, masses, etc. Individuals that are necessarily a part of these masses are not invested with personality and passions; in fact they must be divested of humanity itself. Buckle's view of civilization relies on a very mechanical understanding of humans and their actions. 'That is to say', argues Simpson/Acton, 'he looks at men not as persons, but as machines; and the result he contemplates is not the action of these machines, but their productions'. It is in this way that Buckle is able to deny the fact of free will. By looking at people as masses, as machines, Buckle ignores their individuality, their personality, and by divesting individuals of their personalities they necessarily lose their freedom.[28]

It is argued that Buckle sets up a false dichotomy between free will and fixed laws, that you either accept that man is truly free and ignore the role of laws in society, or you accept fixed laws and abandon a belief in the freedom of thought and action. But 'why make an "alternative" between fixed laws and free-will?'[29] Providence, that Buckle so quickly rejects, is, in Buckle's terms, a fixed law, but it is not of the mechanical sort that would deny freedom.

> Providence dealing with the world is that creative and preservative force which conducts the universe according to 'a law which shall not be broken'; the expression of Providence is this law, wherein no personality can be proved. But Providence dealing with persons is the action of a Personal God upon his personal creatures; warning them, teaching them, judging them. Eliminate personality from your science, and of course your science has nothing to do with the personal providence. ... True liberty is a self-determined, self-chosen perseverance in the way we deliberately think the best. Fixedness, then, is not really opposed to freedom.[30]

One of Buckle's fundamental problems is that instead of proposing fixed laws based on Providential interpretations of the past, he utilizes the so-called science of statistics which does not provide him with the evidence necessary to propose a fixed law.

As the reviewers explain, just because past actions are capable of numeration thereby becoming 'at once a subject of statistics' with 'its average, its maximum, and its minimum' does not mean that the action was not freely chosen. We could attach a number to a variety of actions in the past and inevitably find some regularity concerning the action based on the numbers but this average would tell us nothing about the action itself. Any law derived from such an analysis would simply be a 'law of numbers, a law of chances applicable to numbers and *on the average* applicable to all numerable things; but not implying any force, or cause, or reason why the things themselves should be thus rather than otherwise'. To say that because 'actions are numerable, because they can be averaged, therefore they happened by necessity, by fixed law, is absurd in any man, and in Mr. Buckle dishonest'. This is all to say that by law, Buckle does not actually mean what men of science mean by this term. He actually means 'numerical average'. 'Now it is clear that when a thing has an average, it has an average; you may call this a *fixed law* if you please; but use your terms in such a way that we may not be led into the mistake of concluding that *fixed law* means a necessity inherent in the essence of the thing, and that therefore whatever has an average is necessary, and could not be otherwise.'[31]

The review concludes, unsurprisingly, by suggesting that history, 'on Mr. Buckle's plan, is impossible. For as soon as we simply seek statistics and averages, we have lost sight of man, and are contemplating only his works, his products.'[32] The 'true historian' must take the individual as his subject matter and if he discusses mobs or armies, masses as Buckle would have it, the group must be invested with the same kind of individual personality we find in individuals, with 'wishes, passions, character, will, and conscience'. If it was at all possible to write Buckle's history, which it is not, the reverse would be the case: the historian would have to 'merge the individual in the company, the person in the body; wishes, passions, character, conscience, all would be abstracted ... History would consist in tabular views of births, deaths, marriages, diseases, prices, commerce, and the like; and the historian would be chiefly useful in providing grocers with cheap paper to wrap up butter in.'[33] History under such a programme would be unreadable and 'Mr. Buckle knows better than to reduce history to such dry chaff' which is why the actual narrative of Buckle's history of civilization is not the story of statistics and of the masses. When Buckle 'writes history he makes persons his centres, and reduces it to what it must always be, an intricate and interlacing tissue of biographies ... Thus Louis XIV., Richelieu, and Burke crop out in Mr. Buckle's volume as the centres of his political speculations.'[34]

This simple fact of the matter, the reviewers argue, is that history cannot be 'reduced to a science' on Buckle's plan. The subject matter of history is fundamentally different than that of the physical sciences and it is impossible to generalize man's actions, brought about by his free will, to a singular law. This is the fundamental problem underpinning the philosophy of positivism, a philosophy promoted by 'under-educated, or half-educated men, adepts in physical science, but ignorant of the principles of any other, who insist that all science must have the same methods as theirs, and that metaphysical realities must be measured and explained by physical laws'. There was, according to Simpson and Acton, more than one possible scientific method and therefore more than one way to promote history to the ranks of science. The assumption that the physical sciences provide the only model for gaining scientific knowledge is what was deemed fundamentally wrong with Buckle's work and more generally with the school of philosophy he himself adopted. In this sense, Buckle's 'absurdities and dishonesties are not his own, but those of his school'.[35]

The second review, 'Mr. Buckle's Philosophy of History', begins where the first left off, by admitting that Buckle's primary thesis is problematic in large part because it is the argument of the false philosophy of positivism applied to history. Acton argues that most of the book is simply positivist philosophy masquerading as science, 'completely overgrown and hidden by the mass of matter which is collected to support it, and on which Mr. Buckle has brought to bear all the reading of a lifetime'. Acton has to admit that there is a 'wonderful accumulation of details and extravagance of quotation' but this 'mass of apparent erudition' has the 'manifest purpose of dazzling and blinding his readers'.[36] In other words, Acton does not find Buckle's knowledge worthy of the authority being granted upon him by the excited general reader and he challenges the evidence and authorities Buckle amasses.

First and foremost, Buckle entirely ignores the recent philosophical literature on the progress of civilizations, a surprising absence given the obvious overlap in subject matter. Buckle is also blatantly dismissive of historians in general, in Acton's terms Buckle 'despises the historians', so the question must be asked, 'where, then, are his authorities'?[37] The answer for Acton is quite simple. Buckle relies on the singular authority of Comte, a philosopher who is presented as the 'writer who has done more than any other to raise the standard of history'. 'This is the key to whole book, and in general to Mr. Buckle's state of mind. His view seldom extends beyond the bounds of the system of that philosopher, and he has not sought enlightenment in the study of the great metaphysicians of other schools.' Buckle's knowledge appears to be limited entirely to this author, and when he relies on others, he tends to go to 'the wrong place for information, and ignor[es] the obvious authorities'.[38]

For Acton, this is the problem when you find half-educated positivist philosophers trying to write history. They believe that they can simply apply their method to any field of study without going through the proper laborious training which would involve reading the established authorities. This is precisely what Buckle has not done. 'So far as we have observed, the standard work which is the real and acknowledged authority on each particular subject is never by any chance or oversight consulted for the purpose.'[39] For instance, Buckle's discussion of the history of philosophy is reliant upon 'the most antiquated portion' of Ritter's *Ancient Philosophy*, a 'highly unsatisfactory work'. On the history of medicine Buckle ignores the works of 'Hecker, Häser, and others' in favour of the long displaced works of 'Sprengel and Renouard'. In his discussion of India 'we are referred to a number of obsolete publications, and the great work of Lassen is never mentioned'. On medieval history, Buckle 'shows himself acquainted with just half a dozen of the commonest ... historians'. Acton also goes on at length to describe Buckle's lack of knowledge in the history of Christianity. 'The same ignorance prevails upon almost every branch of learning that is ostentatiously put forward'.[40]

Acton concludes by claiming that the problem with the book, *per se*, has nothing to do with the fact that its author is atheist and promotes secular thinking. 'A book is not necessarily either dangerous or contemptible because it is inspired by hatred of the Church.' Indeed, precisely because it claims to throw 'new light' on a subject matter while increasing our knowledge and attaining a large audience, suggests that the book should be taken seriously. However, Acton believed that he had made Buckle's defects quite clear, even on the book's own terms, and that the book itself enjoyed a readership far beyond any intrinsic merit it might have. Acton worried, however, that perhaps Buckle had debased the discipline of history by presenting a false subject matter. 'In his laborious endeavour to degrade the history of mankind, and of the dealings of God with man, to the level of one of the natural sciences, he has stripped it of its philosophical, of its divine, and even of its human character and interest.' From this perspective, the book's subject matter and method were dangerous, making Acton's review of utmost importance. Thankfully, he was able to make clear the various problems with Buckle's book. 'We may rejoice that the true character of an infidel philosophy has been brought to light by the monstrous and absurd results to which it has led this writer, who has succeeded in extending its principles to the history of civilization only at the sacrifice of every quality which makes a history great.'[41] Acton's message is unavoidable: Henry Thomas Buckle's science of history is not history at all. Nor is it science.

Acton was careful not to suggest that history was not or could not become a science; he was simply clear that history could not become a science along the terms suggested by Buckle. On a few occasions he refers to history as one of the sciences and he mocks Buckle's suggestion 'that before his time there was no science of history'.[42] This position was also mocked in Germany where there was no question about history's scientific status. The discipline of history in Germany was already, after all, referred to as *Geschichtswissenschaft* (historical science) so Buckle's programme of making history a science quite simply did not have the same sort of immediate appeal there that it had in Britain. The German response to Buckle's programme and the nomothetic approach in general was therefore immediate and decisive.[43] In an influential review by Johann Gustav Droysen (1808–84), it was argued that Buckle essentially wanted to make history a natural science (*Naturwissenschaft*) and that this should 'remind us how very unclear, contradictory and beset with arbitrary opinions the foundations of our science are'.[44] He challenged the assumption that seemed to underpin Buckle's work, that there was only one true 'method of knowledge' and that it is the one pursued by the natural sciences. Droysen suggested, instead, 'that if there is to be a science of History, this must have its own method of discovery and relate to its own department of knowledge'.[45] Science (*Wissenschaft*) in Germany was not overtaken by natural science (*Naturwissenschaft*) as was the case in Britain at the same time. It is possible to see in Acton's critique of Buckle as well as in those historians who embraced a science of history but not Buckle's nomothetical approach, a less narrow conception of science or *Wissenschaft* that would embrace empiricism while rejecting the deterministic laws associated with natural science. It is no coincidence that proponents of this approach were clerics who wanted to appropriate the tools of science while maintaining fairly orthodox Christian beliefs that were far too easily overturned in Buckle's world where Providence, free-will and morality played no role in shaping the historical process.

For Acton, Catholics were quite well placed to embrace the benefits of inductive reasoning given that 'they are the only persons who can enter on this field of labour [science] with perfect freedom; for they have a religion perfectly defined, clearly marked off from all other spheres of thought; they alone therefore can enter these spheres free from all suspicion of doubt, and from all fear of discord between faith and knowledge'. This might have been wishful thinking on Acton's part, given the furious debate that was sparked within the Catholic community due to these remarks. However, it was Acton's belief that a Catholic who truly believed in both reason and faith should have absolute 'freedom [to] … move in the sphere of inductive truth'.[46] Buckle's nomothetic method, however, could not grant such freedom.

While Acton's review of Buckle celebrates the slaying of the positivist 'infidel philosophy', privately he was not entirely happy with his review. He felt that he could not quite say what he wanted to say about his own philosophy of history, referring to his review as 'miserably done' and written in 'great disgust and without the smallest satisfaction'.[47] He seemed to articulate a point he was trying to make much better a few years later in a review of Goldwin Smith's *Irish History*. As he argued in this review, it is not the quantity of books and authorities that a historian has read that leads to good history, an argument he clearly makes against Buckle. 'The test of solidity is not the quantity read, but the mode in which the knowledge had been collected and used. Method, not genius, or eloquence, or erudition, makes the historian.'[48] Indeed, the strength of Buckle's book was entirely based on the *apparent* genius of its author, of its eloquence and erudition. It failed as a work of history, however, because it was underpinned by a faulty method which in turn led its author to rely on a whole host of faulty authorities to come to a thesis that simply could not be proven. A better scientific method was necessary, one that would make history, as in Germany, a *Wissenschaft*, and not one that would simply subsume history into *Naturwissenschaft*.

Acton was not alone in his developing conception of history; he was a part of a 'new school' of historians in Britain, a school, in the words of then Regius Professor of Modern History in Oxford Goldwin Smith, 'which requires effect to be produced not by brilliant rhetoric and imposing generalization, but by minute accuracy of detail'.[49] The primary representative of this new school would not be Acton, however, but William Stubbs to whom Acton himself would describe, in 1884, as 'the greatest of all our historians at this present moment'.[50] While Acton would spend the better part of the 1860s and 1870s embroiled in Catholic controversies and then the world of politics, even taking a seat in the British Parliament and becoming Prime Minister Gladstone's chief advisor, Stubbs would seek to guide and direct this new school of British historians.

The appointment of William Stubbs as Regius Professor of Modern and Medieval History at Oxford in 1866 has been viewed as a watershed in the development of history as a professional discipline of study in Britain. He was not from the aristocratic stock of someone like Acton, though his family was certainly among the upper classes. Stubbs was born in Knaresborough, Yorkshire, the son of a solicitor with yeoman ancestry that could be traced back through sixteen generations in Yorkshire. From age seven, Stubbs attended a small private school where he studied Latin, Greek, French, German and Old English, and by the age of fourteen he embarked on a classical education at Ripon grammar school. When his father died in 1842, leaving his mother and six siblings impoverished, Stubbs was lucky enough to secure a servitorship at Christ Church, Oxford, in 1844. Despite the social exclusion such a position entailed, Stubbs received a first-class degree in classics and a third-class in mathematics in 1848. Although Stubbs was raised an evangelical, his religious opinions were heavily informed by the Tractarian movement that was in full swing in Oxford at the time of Stubbs's education. Stubbs considered himself both a High Churchman and a Tory. He was elected Fellow of Trinity College in 1850.

While there was no formal study of medieval history available for students at Oxford in the 1840s, Stubbs spent much time in the Christ Church library pursuing his interests in that field, interests that were earlier cultivated by his father. Stubbs and his father spent much time in Knaresborough Castle pouring over the old medieval records that were preserved there. Stubbs learned, at a very early age, how to read old documents at a time when historians were only just being convinced of the centrality of such historical records to their practice. This early interest in archival material served Stubbs well as he was commissioned to edit the Chronicles and Memorials of Richard I for the Public Record Office, a series of published primary sources later known as the Rolls Series. He engaged in this work from 1863 to 1866, resulting in two volumes of published manuscripts complete with extensive introductions displaying Stubbs's wonderful attention to minute detail.[51]

His appointment to the Regius Chair of Modern and Medieval History at Oxford in 1866 was somewhat surprising given that he was in competition with two much better-known historians, James Anthony Froude and Edward A. Freeman, and Stubbs did not actually put his name up for the appointment. He had felt somewhat dejected for being overlooked for the Chichele Chair of Modern History in 1862 as well as the professorship of ecclesiastical history in 1863, both in Oxford. 'I am not going to stand for anymore [positions]', he wrote to Freeman. 'If I am not worth looking up, I am not ambitious enough to like to be beaten.' He shuddered at the thought of Froude getting the position but he believed that Freeman as Smith's successor would make the most sense.[52] When Stubbs was offered the position, he was just as surprised as Freeman must have been given that he had 'sent in no application'. 'I know, my dear Freeman, that you will rejoice for me as I should have done for you, but I really wish that I could have had my success without your being disappointed.'[53]

At the time, professorial appointments at both Oxford and Cambridge were made by the Prime Minister and parliamentary politics often got in the way of giving the job to the best man available.[54] Political considerations certainly played a role in Stubbs's appointment. Benjamin Disraeli's minority government needed to make an appointment that would appease the Prime Minister's conservative critics, and Disraeli settled on Stubbs because of his Tory politics and conservative religious beliefs. The Reverend Henry Longueville Mansel, Oxford High Churchman, was approached by Lord Carnarvon, who was himself entrusted to make the appointment on behalf of the Prime Minister, and was asked who would be the best choice. It was believed that Mansel was in a position to make such a recommendation as he was, at the time, Waynflete Professor of Metaphysical Philosophy at Oxford. Mansel argued that Stubbs was a 'moderate Conservative' and a 'good Churchman ... whose teaching on religious questions would be thoroughly trustworthy'. Although Mansel did not know Stubbs personally, he felt that 'he is the nearest approach to a Conservative of all the candidates of whom I have heard'.[55]

The political rationale for Stubbs's appointment also went well with his perceived historical abilities. He had not published books as well-known as Freeman's and Froude's but he had edited several volumes of documents for the PRO putting him at the forefront of historians working with primary sources. This likely had less to do with his appointment, however. At the time, the Regius position was viewed as having little influence in actually shaping historical study given the fact that the few professorial chairs had little say in the curriculum. What is more, students were only tested on the material taught by their college tutors who often dissuaded their students from attending Regius lectures that were often seen as a waste of time.[56] Be that as it may, there was much symbolism in the appointment of Regius chairs. Stubbs's appointment in particular was met with subdued joy as well as much hope from the community of historians in Britain that the discipline would finally be taken seriously at one of the ancient universities. This opinion was best expressed in the *Saturday Review*, a weekly journal of politics, literature, science and art.

Considered the most influential periodical of critical opinion of the period, the *Saturday Review* was founded in 1855 by the wealthy amateur A. J. B. Beresford Hope, an eleven-year Conservative M.P. (1841–52) with High Church and even Tractarian leanings. While the weekly paper at times betrayed the religious views of its owner, as was particularly clear in the *Saturday*'s reaction to the supposedly heretical *Essays and Reviews*, it was in theory supposed to avoid theological controversy and was largely beholden to no political party. The journal tended to voice the opinion of its various authors, most of whom were moderate whether Conservative or Liberal. The policy of the *Saturday*'s first editor, John Douglas Cook (1855–66), was to establish a set of competent reviewers and to keep them even if their political or theological views conflicted with the chief proprietor's, which was often the case. This policy, followed by Cook's successors, meant that for extended periods a handful of authors would contribute the bulk of the writing and the journal would therefore reflect those authors' particular views while they remained on staff writing article after article. This was certainly the case when throughout the 1860s and 1870s the *Saturday Review* became the mouthpiece for a professionalized history under the guise of an inductive methodology largely because Freeman and J. R. Green authored most articles on historical subjects during that period.[57] It should not have been a surprise, then, when the *Saturday* believed that Stubbs's appointment was a great victory for the discipline of history. It celebrated in its typically sarcastic tone that 'the powers that be' have finally admitted that the study of history actually 'exists'.[58]

The *Saturday Review* came out in favour of the appointment following Stubbs's inaugural lecture at Oxford on 7 February 1867 where Stubbs made it clear that he wanted to be 'instrumental ... in the founding of an historical school in England', one that would promote and teach a kind of history that would be diametrically opposed to the philosophically based history found in the *History of Civilization in England*.[59] Indeed, Stubbs introduced himself in a telling way, 'not as a philosopher nor as a politician, but as a worker of history'.[60] Stubbs made it clear that he was not going to use his position to make 'proselytes to one system' but that he would use his 'office as a teacher of facts and of the right habit of using them'.[61] Stubbs wanted to break from the notion of the philosopher historian, the genius, and he presented himself as a worker, a worker of history whose task it was to train other workers of history.

For Stubbs, history was not complicated. One need not be a genius to engage in historical analysis which is simply 'the process of acquisition of a stock of facts'.[62] While this process is not complicated, it is by no means easy. It requires, at times, painstaking concentration and is often tedious and can be unrewarding. He admitted that Buckle was right about one thing, that the recently discovered facts of the past are increasing at an astounding rate due to the opening and expansion of archives and he warned that the 'masses of information ... are threatening to overwhelm us'.[63] In order to avoid being overwhelmed, he argued that historians could not work away on their own with their own particular methods and their own particular prejudices. For history to benefit from the 'fountains of historical refreshments' there would need to be, Stubbs argued, 'a great republic of workers'.

> Happily, there is no lack of helpers; the great German hive of historical workers is busy as we are on our archives: such and so close are the ties which now, owing to the facilities of travelling and communication, the abundance of libraries and the accessibility of records, the extension of literary and investigative sympathies, and, I am happy to think, the extinction of literary jealousies, are now binding the historical scholars of Europe, that I think and hope that the day is coming when, although we may not cease to quarrel with and criticise one another, there will be a great republic of workers able and willing to assist one another; not working for party purposes unfettered by political prejudices, and although as strong partisans and politicians as ever, anxious above all to find the truth, and to purify the cause that each loves best from every taint of falsehood, every inclination to calumny or concealment.[64]

This was a grand vision for the burgeoning historical community, one that would transcend national boundaries and be linked together by a common method. Stubbs spoke of the need to 'join with the other workers of Europe in a common task'. That common task will 'build, not upon [Henry] Hallam and [Francis] Palgrave and [John] Kemble and [James Anthony] Froude and [T. B.] Macaulay', in other words, antiquarians, popularizers and Romantics, 'but on the abundant collected and arranged materials on which those writers tried to build whilst they were scanty and scattered and in disorder'.[65] We will not build, in other words, on the names of the authors, but on the history that they collected on their own without the help of a community of fellow workers or newly catalogued archives. Three years before his appointment, Stubbs wrote to Freeman that he 'should like to see [Lord Bacon] *rehabilitated*' given the way in which historians had been ignoring the important foundation provided by a wealth of minute facts.[66] His inaugural publicized that wish.

The importance of the lecture did not go unnoticed by those in attendance. In a letter to Freeman, J. R. Green proclaimed that 'Stubbs piped unto me and I danced'. Greene could not help but feel the 'religious glow' of the lecture as well.[67] Indeed, Stubbs seemed to suggest that his proposed method of study would find much common ground with natural theology. History, Stubbs argued, should be studied, 'in common with Natural Science, in the study of living History' in order to perceive 'the workings of the Almighty Ruler of the world'.[68] It should not be surprising that Anglican historians, in particular, looked to natural theology as providing a model for engaging in a scientific pursuit while avoiding any conflict with Christian dogma. For natural theologians such as William Paley, one studied nature in order to uncover God's Providential design. What was more, nature provided proof of that Providential design; nature's beautiful order and symmetry is evidence of design and therefore of a designer. This method taught that naturalists should really only concern themselves with questions of secondary causes and limit their analysis to descriptions of how the world functions while leaving questions concerning first causes, that is, the origin of things, to the theologians.[69] And so it would be for history under Stubbs's conception as well as most other historians adopting the inductive approach.[70] This appeared to have little effect on their actual scholarship because, as Stephen Bann has argued, 'the notion of Providence enabled the diligent researchers to forego controversy, in the confidence that any particular nugget of fact would be compatible with a divine purpose which could never be revealed as a whole'.[71] But such a view certainly entailed rejecting any historical theory that would disrupt Providence.

This gets at the heart of why Stubbs became such a representative for the distrust of the 'philosophy of history', a discipline, he argued, 'which exalts the generalizations of partially informed men into laws, and attempts out of those laws to create a science of history'. Such a school he was happy to reject on the basis of his belief in Divine Providence, which, 'by its very nature acts with some uniformity of cause and consequence'.

> But I also believe that the Divine Providence acts in the government of the world through secondary agencies, and the chief agency in the department of history which we are attempting is the will of men, the aggregate of wills of individual men ... It is the purity of history to trace the workings of these secondary agencies, and even to generalize for them; but to enter into the higher regions of Divine Providence is the portion of faith rather than of science, and I for my part should be very loath to bind as by a law of action of Divine Providence with any generalization of mine from men's doings, as regards either past or future.[72]

In other words, for Stubbs as well as for Acton and other inductive historians, Divine Providence should not be the historian's direct object of study; however, the historian must not engage in the kind of generalizing that could contradict Providence; 'perfect knowledge is independent of, and even inconsistent with, any generalization at all'.[73] The inductive science of Ranke, because of its focus on the building of individual facts could easily be incorporated into such a view, whereas the supposed 'positivist' approach of Buckle could not.

In 1873 the first volume of Stubbs's *Constitutional History of England* would appear becoming a profound symbol of the inductive science of history that he promoted in his famous inaugural lecture at Oxford. His voice intrudes into the narrative of Britain's constitutional development only twice, in the preface where he explains his method and in the conclusion to the third volume where he argues that the purpose of the book was 'to train the judgment of his readers to discern the balance of truth and reality'.[74] The 'reality' of English history was presented as a glacially slow unfolding of the origins and development of the main structures of England's political life. While Stubbs would have rejected the suggestion out of hand, his narrative reads very much like a natural history in the Darwinian mode and not just in the very deliberate plot but in the organic way in which Teutonic institutions are transferred to England and adapted and further developed. The *Constitutional History* is a developmental narrative that certainly reminds the reader of a Darwinian evolutionary approach where change occurs but slowly and not out of context with the surroundings. Stubbs was more likely channelling Burkean traditionalism than Darwinian evolution, however, and yet both approaches appear as kindred spirits in the *Constitutional History*, as proponents of the same basic inductive approach that Buckle so thoroughly rejected as a mere collection of facts. Stubbs showed that such a collection could be presented in a way that would explain the centrality of tradition *and* adaptation in Britain's political development much better than an analysis that foregrounded laws ever could.[75] Indeed, Buckle's much more mechanical approach seemed to highlight long periods of stasis followed by sudden changes in the same way that catastrophists considered the earth's history.

If Stubbs was outspoken in his rejection of any overarching philosophy of history, of any deterministic science, it is impossible to read even the first few pages of Stubbs and ignore the centrality of scientific racism that infiltrates his pages. As he explains in those early pages, his study was entirely based on answers to a series of preliminary questions: 'Who were our forefathers, whence did they come, what did they bring with them, what did they find on their arrival, how far did the process of migration and settlement affect their own development, and in what measure was it indebted to the character and previous history of the land they colonised?' In answering these questions, Stubbs gave an explanation that in a sense explained the entire argument of his book, of the centrality of continuity and slow adaptation in England's political history.

> The English are not aboriginal, that is, they are not identical with the race that occupied their home at the dawn of history. They are a people of German descent in the main constituents of blood, character, and language, but most especially, in connexion with our subject, in the possession of the elements of primitive German civilisation and the common germs of German institutions. This descent is not a matter of inference. It is a recorded fact of history, which these characteristics bear out to the fullest degree of certainty.[76]

Stubbs was here relying on the new science of race that helped provide him with the foundational facts for his study, that the English were of German descent and had long ago embraced primitive Germanic institutions. The English, because of the peculiarity of their Island nation, were able to purify this dual inheritance in a way that was impossible on the Continent, becoming more German than the peoples of Germany. 'Language, law, custom and religion preserve their original conformation and colouring. The Germanic element is the paternal element in our system, natural and political.'[77]

The centrality of race in explaining the development of English history, for Stubbs, was another aspect of his own anti-philosophy of history in diametrical opposition to that of Buckle. Buckle had tried to dissuade historians from being charmed by the race concept in his *History of Civilization*, arguing that 'original distinctions of race are altogether hypothetical', entirely outside the realm of scientific truth, while those distinctions 'which are caused by differences of climate, food, and soil' are central in shaping a nation's history. Buckle, rather, adhered to the opinion of John Stuart Mill, 'one of the greatest thinkers of our time', who argued that 'of all the vulgar modes of escaping from the consideration of the effect of social and moral influences on the human mind, the most vulgar is that of attributing the diversities of conduct and character to inherent natural differences'. Buckle went on to criticize those '[o]rdinary writers [who] are constantly falling into the error of assuming the existence of this or that difference; which may or may not exist, but which most assuredly has never been proved.'[78] For Stubbs, however, there was scientific proof, not just of the existence of race, but of the particular Germanic racial inheritances of the English nation – both its people and politics. What was more, argued Stubbs, the history of England's constitutional development was further proof of this racial inheritance: 'the chain of proof is to be found in the progressive persistent development of English constitutional history from the primeval polity of the common fatherland.'[79]

In this regard, Stubbs's work reflected the growing centrality of race in both scientific and historical discourses of the mid-Victorian period. Of particular note was John Kemble's *The Saxons in England* (1849), a work that analysed early Saxon forms of organization that were said to have brought to maturity even earlier Teutonic settlements thereby providing the foundation for England's racial and imperial greatness. Kemble had studied Teutonic philology under Jacob Grimm. Like Grimm, Keble believed one could trace racial descent through language but it was the philologist Max Müller who established the links between an original Aryan language and people from which the English derived. His series of lectures at the Royal Institution on 'The Science of Language' (1859–61) made explicit the connections between the early Teutonic and Saxon foundations of English civilization and later Anglo-Saxon superiority.[80] He would much later in his life back away from the explicit identification of race and language but not before historians such as Stubbs and especially Edward A. Freeman embraced philology as providing evidence of linguistic and therefore racial continuity in English history.[81]

The continuity of English history, both racial and institutional, was the primary focus of Freeman's medieval studies, most notably his multi-volume *History of the Norman Conquest* (1867–76). If any event threatened the continuity of English history, it was the conquest that saw England ruled by Norman invaders, a process that began in 1066. Freeman spent five thick octavo volumes explaining how this conquest, rather than representing a fundamental transformation in English history as had been the usual interpretation, was a mere moment of disturbance quickly absorbed by the long march of English liberty. The reason so many previous historians got it wrong, argued Freeman, was because they began their studies at 1066. 'They thus failed to perceive that the Norman Conquest, instead of wiping out the race, the laws, or the language which existed before it, did but communicate to us a certain foreign infusion in all three branches, which was speedily absorbed and assimilated into the preexisting mass.'[82] Even though the conquest brought about a 'foreign infusion, which affected our blood, our language, our laws, our arts ... the older and stronger elements still survived and in the long run they again made good their supremacy.'[83] This was, much like Stubbs's much longer history of the English constitution, a story of continuity and adaptation. It was also a story of race.

Stubbs and Freeman shared this interpretation of English history in large part because they were such close friends who also shared a common background and educational history. Freeman was Stubbs's senior by two years and they would also both be elected fellows of Trinity College, Oxford, Freeman five years before Stubbs in 1845. Also like Stubbs, Freeman was a High Church Anglican and was heavily influenced by the Tractarian movement that had infiltrated Oxford's theological life throughout the 1840s. He was not terribly concerned with issues of theology, however. He was more interested in church history itself. This early embrace of ecclesiology meant that Freeman was highly influenced by the antiquarian pedantry that tended to be a hallmark of ecclesiological societies such as ensuring correct dates, nomenclature and spelling.[84] Such must account for the almost fanatical devotion Freeman would later exhibit towards absolute historical 'accuracy' not just of the facts but of proper spelling and language. Indeed, it was the antiquarian and ecclesiologist in Freeman rather than any reading of Rankean or Baconian prescriptions – as was the case for Stubbs – that largely account for his inductive tendencies, though he certainly learned to present them within the more fashionable Rankean rhetoric. Freeman in particular hated the way the English language was being corrupted by foreign influences and he was always sure to restore the 'genuine spelling' of Old English and Greek names in his studies refusing to perpetuate the 'chaos of French and Latin corruptions.'[85] Language was intimately connected with race, for Freeman. Corrupting the English language meant corrupting the English race and he would not allow his own studies to contribute to such racial degeneration.

Indeed, historical 'purity' rather than 'accuracy' might better explain Freeman's methodological mantra. The Teutonic racial purity of the English nation was, much as in Stubbs's portrayal, the central rationale in favour of his thesis about the absorption of the Norman invaders into 'our national being'. According to Freeman, William the Conqueror and his Norman successors for the most part accepted the English constitution. Certainly changes were made that apparently transformed the nation but within only 'a few generations we led captive our conquerors'.[86] It helped that the Normans were not an entirely 'foreign' invader. They were not racially French, according to Freeman, but Teutonic and could therefore be racially assimilated by the Anglo-Saxons. 'The conquered did not become Normans, but the conquerors did become Englishmen.'[87]

While Freeman would become one of the loudest proponents of an inductive science of history as well as a proponent for the use of the philological comparative method in historical analysis of races and languages,[88] he agreed with Stubbs's and Acton's distaste for Buckle's particular science of history. In one of his important essays originally appearing in the *Edinburgh Review* in 1860, Freeman gave a broad sweep of the 'Continuity of English History'. It was England's continuity, particularly 'the steady course of freedom', argued Freeman, that separated English history from that of both France and Germany. However, Freeman explained, there was no 'universal formula' to explain this central fact of English history. He had to 'confess', therefore, that he was 'not up to the last lights of the age' and that he had 'not graduated in the school of Mr. Buckle'. He did not think it was right to 'submit the phenomena of English history, its course at home or its points of difference from that of other nations, to any grand scientific law'. Freeman was adamant that he and English historians must 'retain our faith in the existence and the free-will both of God and man' something that the school of Mr. Buckle clearly rejected. Only by studying the peculiarities of English history, of its 'national character, geographical position, earlier historical events' and the 'personal character of individual men', will the historian be able to understand England's 'uninterrupted historical continuity'.[89]

It was against Buckle's science of history that Acton, Stubbs and Freeman articulated their own brand that left a space for Providence, free will and, in the case of Stubbs and Freeman, scientific racism. This was a method that would foreground the minutia in order to establish a proper foundation on which a true understanding of history could be built. Many other English historians would come to share this view, most notably J. R. Seeley. He believed, in particular, that Buckle's project with its focus on the masses and on averages was not history at all.[90] Stubbs's initial knee-jerk opinion about Buckle's science of history would find resonance after all, despite his own concerns that he was committing heresy by rejecting it. However, it was not the only response. While one group of historians used Buckle's work as a foil to promote their own science of history, another more romantically inclined group of historical writers criticized Buckle's work in a way that challenged the very possibility of a science of history.

3 CONTROVERSIAL BOYS

These lectures will not be, in the popular sense, history at all.

Charles Kingsley, *The Roman and the Teuton* (1864)

But the drama of history is imperishable ...

James Anthony Froude, 'The Science of History' (1864)

While Stubbs was attempting to found a school for historical workers under the new inductive methodology at Oxford, his counterpart at Cambridge, Charles Kingsley (1819–75), was far less interested in advancing the cause of history as an autonomous and scientific discipline of study, or so believed his critics. Stubbs and Kingsley could not have been more different and not just in historical outlook. Stubbs was, as we have seen, a High Churchman and Tory, whereas Kingsley's Anglicanism was of the *Essays and Reviews* variety. He was also, for much of his life, a Christian socialist. Where controversy seemed to follow Kingsley throughout his life, Stubbs avoided it at all cost, believing it to be the natural result of an inherent subjectivity that should be suppressed. 'Of all things in the world except a controversial woman', Stubbs said at one of his Oxford lectures, 'a controversial boy is the most disagreeable'.[1] Stubbs was also a self-described historian, an editor, a diligent archival worker. Kingsley, on the other hand, would likely have never referred to himself as a historian: at best he was an historical novelist; at worst he was a writer of autobiography who hid his opinions behind the veil of 'fiction'. Most problematic was Kingsley's rejection of scientific history in favour of a more subjective and artistic method in diametrical opposition to that of Stubbs and the nascent historical profession. But Kingsley was not alone.

Kingsley was appointed Regius Professor of Modern History at Cambridge University in 1860. The appointment was not made without controversy. The novels Kingsley had become famous for writing by that time helped to secure his candidacy for the position but they also made him a polarizing figure, especially at Anglican Cambridge. Kingsley's novels, while historical, were primarily influenced by the religious controversies that rocked the Church of England in the 1840s and 1850s. Kingsley was deeply concerned about the return to Catholic Anglicanism advocated by a growing faction of Oxford scholars. The Oxford Movement, as it was called, was led by John Henry Newman, Edward Pusey and John Keble. They and their supporters were often referred to as Tractarians for the series of publications they produced, the so-called Tracts for the Times that sought to re-establish the autonomy of the Church of England in the face of liberal reform. The main thrust of the Tractarians' doctrinal arguments was that the Church needed to accept its Catholic roots and they promoted a series of reforms that would see the Church revoke many of the Protestant practices that had been implemented since the era of Henry VIII.[2] Newman, for instance, in *Tract 90* (1841), argued for the necessity of celibacy among Church clerics. Kingsley was particularly irritated at the way in which Tractarians celebrated the lives of celibate Catholic monks and he shot back in his historical novel *Hypatia* (1853) by writing about the fanatical monks of fifth-century Alexandria who murdered the philosopher Hypatia for her Neoplatonic views. *New Foes with an Old Face* was the book's not-so-subtle subtitle.[3] Many saw Kingsley as heroically defending the religious national institution against those who wanted to reject the English Reformation and once again surrender the religious conscience of the nation to the authority of Rome. Queen Victoria and the prince consort certainly shared this view and this should go some way towards explaining why he was appointed Regius professor in 1860. However, many others, while rejecting the popery promoted by some Tractarians, could still empathize with the need to strengthen the Church of England as an autonomous institution that could be seen as slowly but surely coming under the control and whims of a democratizing state apparatus. From such a perspective, Kingsley's attacks on Tractarians were seen as attacks on the traditions and power of the Church itself. It did not help matters that as Kingsley shot back at the likes of Newman, he also criticized some of the remnant Catholic practices of the Anglican faith, and many at Cambridge did not approve.

Of particular concern for Cambridge elites were Kingsley's remarks about Cambridge itself and the ancient university system as a whole contained in his novel *Alton Locke* (1850). Kingsley had attended Magdalene College of Cambridge in the early 1840s and he hated the blatant favouritism shown to students of a well-bred stock, a favouritism that was not shown to the young man from a lower class than most of his colleagues. He refused to play the patronage game that in part involved giving the dons gifts and hiring expensive tutors to ensure good grades, a system of education that could only benefit the rich at the expense of the poor and lower to middle classes. 'It is a system of humbug, from one end to the other' the initially naive hero of *Alton Locke* is informed by his cousin about the ancient universities.[4] 'Can't you see? The whole is monastic-dress, unmarried fellows, the very names of the colleges. I daresay it did very well for the poor scholars in the middle ages, who, three-fourths of them, turned either monks or priests; but it won't do for the young gentlemen of the 19th century.' Kingsley was very critical of a university system that refused to reform itself in the face of a society radically different than the one that gave birth to it. Centuries of stagnation have now led to a situation where Cambridge elites 'are afraid to alter anything, for fear of bringing the whole rotten old house down about the ears. ...That's why they retain statutes that can't be observed; because they know, if they once began altering the statutes the least, the world would find out how they have themselves been breaking the statutes. That's why they keep up the farce of swearing to the Thirty-Nine Articles, and all that; just because they know, if they attempted to alter the letter of the old forms, it would come out, that half the young men of the university don't believe three words of them.' Because of this the ancient dogmas are followed to the letter but the spirit, the Christian spirit that led to the creation of the once noble universities, is thrown away. Only those who will 'sign the dogmas of the Church of England' are admitted while all others are excluded for the simple reason that they are poor.[5] The ancient universities were all Catholic dogma without the Protestant spirit.

While Kingsley backed away from much of the radical Chartism of his youth, he embraced a form of Protestant Christianity that was seen by many as quite dangerous and possibly harmful to Anglican doctrine. Kingsley would become a symbol of the 'Broad Church', a largely pejorative term used primarily by High Church Anglicans to group together a diversity of liberal-minded Anglican clerics who advocated reforming old church (read: Catholic) doctrine in favour of a Christianity more open to scientific advances, particularly where such advances seemed to present new challenges to the Christian faith.[6] Biblical criticism, uniformitarian geology and the new science of evolutionary biology seemed to challenge old Christian dogmas, from central facts in the life of Jesus Christ and the history of Creation to the very nature of existence. Kingsley argued that such challenges needed to be faced head on and he believed that when it came down to it the fundamental truths of the Christian faith could only be strengthened by embracing the new scientific theories. Charles Darwin most famously quoted from Kingsley as an unnamed 'celebrated author and divine' in the second and all subsequent editions of the *Origin* as arguing that 'it is just as noble to a conception of the Deity to believe that He created a few original forms capable of self-development into other and needful forms, as to believe that He required a fresh act of creation to supply the voids caused by the action of His laws'.[7] Despite the fact that Darwin had managed to provide the causal mechanism for evolution, there was still something deeply mysterious about that process. That inherent mystery and wonder at the core of evolution greatly attracted Kingsley as evidence of God's central role in the process. Such views, however, were certainly not orthodox among Anglican clergymen, and when a handful of like-minded clergymen published a series of essays which drew on the new sciences while rejecting biblical literalism entitled *Essays and Reviews* (1860), they were viciously attacked by the Bishop of Oxford and others. A formal trial of heresy ensued, though it was largely unsuccessful.[8]

What does all this have to do with Kingsley's views of history and historical methodology? By the time of his appointment Kingsley had not written anything of substance about the discipline of history, though his historical novels do contain some germs of his historical perspective that would become more clear during his tenure at Cambridge. Kingsley, it is often said, was a 'muscular Christian', a term coined by T. C. Sandars in his review of Kingsley's *Two Years Ago* (1857) for the *Saturday Review*.[9] The protagonists of Kingsley's novels tended to be viral, manly men, overcoming their problems and constraints through heroic struggle and generally the embrace of a Protestant Christianity. This perhaps placed him within the Thomas Carlyle school of great man history, though his devotion to such a perspective would be made much more clear in his inaugural lecture at Cambridge. As 12 November 1860 approached, the anticipated date of Kingsley's inaugural lecture, much apprehension was in the air about his appointment. The weekly *Literary Gazette* reported: 'That the foremost apostle of muscular Christianity should be selected to instruct the youthful members of this august society in a subject so wide and so capable of arbitrary interpretation as history, was regarded with grave apprehension by one party, with ridicule by another, and with unmingled satisfaction by a third.'[10] Most historians, it is safe to say, were among the group that viewed the appointment with ridicule. 'What a horrid appointment of Kingsley!' Stubbs wrote to Freeman. 'I suppose that it is on the principle of putting the worst man in the best place, so that you have all the good ones trying to show how much better they are, and so benefiting the world.'[11] Freeman spoke for many when he said that the appointment 'seemed ... an inexplicable freak' and 'a joke'.[12] '[O]f all the men living he is the least qualified to undertake the work of an historian or an historical Profession.'[13]

Despite Kingsley's lack of historical training he did seem to read the mood of the profession correctly when he decided to make the subject of his talk 'The Limits of Exact Science as Applied to History'. Without ever mentioning Buckle by name, the lecture was meant to be a direct assault on the 'positivist' science of history. In its place Kingsley proposed a method of history that accepted the mysterious and indeterminate nature of the past through an analysis of great men – those historical aberrations who refuse to conform to the law of averages.

The lecture was given in the Senate House that was apparently packed full with critics and supporters alike wanting to hear one of Kingsley's sermons. He began by setting expectations fairly low with the rather unremarkable statement that he would 'endeavour to teach Modern History after a method which shall give satisfaction to the Rulers of this University'.[14] Kingsley was clearly not promising to teach history under some new and exciting historical methodology or found a new school of history as Stubbs would six years later at Oxford. No; Kingsley's lecture was more of the 'how not to write history' approach than of the 'how to', yet he provided a method nonetheless, one that established a fair amount of continuity between himself and previously popular English historians.

Those in the audience concerned that Kingsley's supposed Broad Church embrace of contemporary scientific fashions would thrust him into the positivist school of history were pleasantly surprised. Kingsley admitted that there was little doubt that 'History obeys, and always has obeyed, in the long run, certain laws' but those laws he explained 'are to be discovered, not in things, but in persons; in the actions of human beings'. We will only understand historical laws, argued Kingsley, when we come to understand the actions of people. He was deeply concerned that the current fashion of rushing in and applying positivistic science to all avenues of human knowledge was actually working to undermine that knowledge. '[T]he rapid progress of science', argued Kingsley, 'is tempting us to look at human beings rather as things than as persons, and as abstractions (under the name of laws) rather as persons than as things. Discovering, to our just delight, order and law all around us, in a thousand events which seemed to our fathers fortuitous and arbitrary, we are dazzled just now by the magnificent prospect opening before us, and fall, too often, into more than one serious mistake.'[15]

The serious mistakes arise from working backwards from the laws to the facts, often by making all the facts conform 'to the very few laws' known forcing specifically moral or spiritual phenomena under the umbrella of laws meant to explain the physical or economic world. Kingsley was of course challenging Buckle's assertion that the moral and spiritual were entities derived from material circumstances, 'a hasty corollary ... and one not likely to find favour in this University'. Kingsley went on to cheekily dismiss one of Buckle's more conspicuous arguments: 'We shall not be inclined here, I trust, to explain (as some one tried to do lately) the Crusades by a hypothesis of overstocked labour-markets on the Continent.'[16]

Kingsley was not suggesting that there were no laws to explain a diversity of phenomena, just that such laws 'are beyond us'. There will always be exceptions that disprove rules and just when historians believe they have found a general law to contain the diversity of facts a great man will appear and break the supposed indestructible laws of nature. '[G]reat nature, just as we fancy we have found out her secret, will smile in our faces as she brings into the world a man, the like of whom we have never seen, and cannot explain, define, classify – in one word, a genius.' The rise of great men, of geniuses, cause the historical process to take 'quite new and unexpected paths, and for good or evil, leave their stamp upon whole generations and races'.[17]

For Kingsley it was Buckle's error to suggest that 'the history of mankind is the history of the masses' and that it was likely more accurate to say 'that the history of mankind is the history of great men'. If we were going to establish history on a scientific footing, it would be in accepting the way in which great men shape history; and if we were actually to discover the laws that underpinned historical development it would be in analysing the way in which 'great minds have been produced into the world, as necessary results, each in his place and time'. 'That would be a science indeed', argued Kingsley, but such a science is certainly in an infant state. 'As yet, the appearance of great minds is as inexplicable to us as if they had dropped among us from another planet.'[18]

While Kingsley seemed to be proposing a science of great men to take the place of Buckle's science of the masses, he set the bar extremely high in order to live up to the terms of this new science. In order to understand a great man such as Luther, for instance, in order to understand how Luther was able 'to have been a person with an originally different character from all others', it is first and foremost necessary to 'find out what he was'. This, for Kingsley, was no simple task. Much easier was the task set out by positivists who 'settle beforehand our theory, and explain by it such parts of Luther as will fit it'.[19] More difficult, and much more profitable, would be to 'learn the laws which produced Luther, by learning Luther himself; by analysing his whole character; by gauging his powers; and that ... we cannot do till we are more than Luther himself.' To understand truly Luther's character, then, Luther the man, the historian must be greater than Luther himself, 'because to comprehend him thoroughly, he must be able to judge the man's failings as well as his excellencies; to see not only why he did what he did, but why he did not do more: in a word, he must be nearer than his object is to the ideal man'. In other words, in order to achieve a science of great men, the historian must be a great man himself, a genius, in order to comprehend that most 'mysterious element of all character, which we call strength'.[20]

Following Kingsley's criteria of understanding great men would be virtually impossible. This was, I think, precisely Kingsley's point. There is something both 'disturbing' and 'miraculous' at work when great men appear and undermine the previously careful calculations of scientific understanding. The one constant in human history, according to Kingsley, is its unpredictability. 'So far removed is the sequence of human history from any thing which we can call irresistible or inevitable.' Kingsley argued that he was not trying 'to discourage inductive thought' but he was adamant that the hope be 'given up, at least for the present, of forming any exact science of history'.[21]

Whether a science of history is possible or not, Kingsley's ultimate message was that such a conception was largely beside the point. If students and historians want to 'understand History, they must first understand men and women. For History is the history of men and women, and of nothing else; and he who knows men and women thoroughly will best understand the past work of the world, and be best able to carry on its work now.'[22] The evidence for such an understanding is not found in government documents, in archival traces, but in biographies, in the study of individuals and even human nature. 'If, therefore, any of you should ask me how to study history, I should answer – Take by all means biographies: wheresoever possible, autobiographies; and study them. Fill your minds with live human figures; men of like passions with yourselves; see how each lived and worked in the time and place in which God put him.' Historians need to bring the dead back 'to life again' and 'see with his eyes and feel with his heart'. It is only in this way that 'you will begin to understand more of his generation and his circumstances, than all the mere history-books of the period would teach you. And not only to understand, but to remember.'[23] It is not the innumerable facts of the period that one must build into some monument of order to understand the past; it is in understanding the individuals, the biographies and autobiographies, where true historical understanding is achieved.

It is true that many in attendance found Kingsley's lecture full of insight and given in a powerfully effective rhetorical style that was captivating. The nonconformist *British Quarterly Review*, for instance, found the lecture to be 'an admirable production, and promises well'. The reviewer argued that Kingsley was 'a man far too healthy in soul to be a disciple of Mr. Buckle' and was particularly pleased at Kingsley's 'onslaught on the narrow and heartless dogma which would make the science of history to be only another form of the science of statistics'.[24]

If the periodical press as a whole is to be taken as an indication of the reception of Kingsley's lecture, however, many more in attendance found his foray into the intricate nuances of history's scientific methodologies hopelessly confused and confusing. The *Literary Gazette* found the tone of Kingsley's lecture quite rightly 'unquestionably negative' but it was Kingsley's general argument that the *Gazette* found most troublesome. 'Our readers must be forcibly struck with Mr. Kingsley's inconsistence in first of all attempting to demonstrate the impossibility of an exact science of history by attempting to demonstrate that historical sequences are not subject to law; next by admitting that such a science may at some more advanced stage of the human understanding be essayed; and finally, with a shortsightedness almost inconceivable in such a man, impotently depreciating any pursuit after that science, "at least for the present."'[25] In a theme that would haunt Kingsley's reign as Regius professor, the *Gazette* concluded that Kingsley has simply 'not given himself time enough to prepare his premises or mature his conclusions'.[26]

The *London Review*, a weekly journal of politics, literature, art and society, was similarly concerned about the title of Kingsley's lecture, a title that 'arouses suspicion that he has not mastered the subject he ambitiously and unnecessarily discusses, nor reflected very deeply on the difference and distinction between exact science and history'. Upon reading the printed lecture, the *London Review* confirmed that Kingsley had 'hastily snatched up a subject, and resolved to lecture on it, without previous knowledge or due consideration'.[27] The reviewer suggested that perhaps Kingsley was right, that men of genius certainly play a role in the historical process, though the extent to which such was the case was far less than Kingsley was willing to admit. Most damagingly, if Kingsley's method was to be followed, it was argued that such would 'carry back the youth of Cambridge and the study of history, to the singularities, the crotchets and the extravagances of individuals. He would substitute exceptions for rules, and ancient barbarism, with its rude caprices and passions, for modern refinement and modern regularity.'[28] The free-thinking *Westminster Review* also criticized the lecture as being simply confusing: confusing in the way in which Kingsley sought to blur the very clear boundaries between history and biography; and confusing in the way in which Kingsley used the term science in relation to history throughout the lecture.[29]

It is somewhat surprising that the *Westminster Review*, a periodical generally sympathetic to Kingsley's more liberal views, would criticise Kingsley so heavily. And yet, the reviewer suggested that undergraduates should 'not measure their Professor by his inaugural lecture', that he would likely win them over by his enthusiasm for the past and his wonderful storytelling capabilities.[30] The *Westminster Review* was certainly right that Kingsley's lectures would be very popular with the students as he lectured to crammed halls on a regular basis during his professorship. Unfortunately Kingsley's reputation as an interloper did not improve with time. When a series of his lectures were published in 1864 entitled the *Roman and the Teuton*, he was viciously attacked from several quarters and this time not just for hastily speaking on a specific subject; now he was being attacked for teaching on behalf of an entire discipline whose basic methods, procedures and facts seemed beyond his comprehension.

Largely at issue with Kingsley's lectures was precisely the fact that he seemed more interested in telling a good story, in letting students get a feeling about the past and in making friends with the dead than in providing his students with a training in the study of particular facts. Edward Freeman was utterly beside himself at the way in which Kingsley brought the historian 'down to the level of sensation novelist'. Freeman was willing to admit that Kingsley could certainly tell a good story but he felt that Kingsley was simply devoid of the 'requisite learning' as well as the 'requisite state of mind'. It was clear from the lectures that Kingsley set about to gain the learning he had been missing, but there was clearly an 'inherent defect of the mind' that was 'unconquerable'. 'There is a natural twist about his mind, as about his style, which effectually incapacitates him as a teacher and a learner of history.'³¹ He thought it would have been impossible, had he not read it for himself in the *Roman and the Teuton*, 'that a Professor in an English University could begin a series of historical lectures by a parable about Trolls and Forest children'. He was referring to Kingsley's lecture 'The Forest Children', which was ostensibly about the destruction of Rome. By this time Kingsley had written and published a few books for children, most notably *The Water-Babies* (1863), a satiric commentary on current affairs from scientific controversies to politics and education. Freeman believed that 'The Forest Children' was perhaps a story 'pretty enough to put in a penny book and give to a "land-baby," and, for aught we know, it may be suited to the intellect of poetic tailors and philosophic gamekeepers, but it is simply insulting to put forth stuff of this sort to an assembly of academic hearers of any age'.³² The lectures were proof enough that Kingsley was no historian but merely a sensational novelist better suited to teach children than undergraduates about the finer points of history. This view was widely shared.

Even the monthly *Eclectic Review*, a journal founded in 1805 and edited by Dissenters quite sympathetic to Kingsley's literary productions, found the *Roman and the Teuton* to be 'about as thoroughly incompetent a book upon such a subject as could be produced by a really able man'.³³ The general judgement about his inaugural lecture held true for these lectures as well, lectures that were 'evidently the result of haste, of reading, the accumulation rather of simple pleasure than of analysis and labour'. The reviewer admitted that those appreciative of Kingsley's historical fiction will read the present volume 'with great delight' as it is written 'after the same *Impulsia Gushington* kind of style'.³⁴ Not only did Kingsley's historical work not bear out the painstaking labour that was so central to the work of someone like Stubbs, it most problematically followed the style of fiction writing, a discipline most historians were seeking to demarcate from the much more serious discipline of history. Kingsley himself admitted on the first page that '[t]hese lectures will not be, in the popular sense, history at all'.³⁵

This view of Kingsley's lack of historical competence stuck. When a second edition of the *Roman and the Teuton* was published in 1891, Kingsley's friend the philologist Max Müller penned a preface that tried to explain why reading the book was still useful despite the fact that much of the criticism directed against it was justified. Müller explained that he was not so 'blinded' by his friendship with 'Kinsley as to say that these lectures are throughout what academic lectures ought to be'. He acknowledged that 'they do not profess to contain the results of long continued research' and that they 'are not based on a critical appreciation of the authorities which had been consulted'. What was perhaps worse, Müller explained, was that 'these lectures were not always written in a perfectly impartial and judicial spirit, and that occasionally they are unjust to the historians who, from no other motive but a sincere regard for truth, thought it their duty to withhold their assent from many of the commonly received statements of mediaeval chroniclers'.[36] He admitted quite openly that '[h]istorians by profession would naturally be incensed at some portions of this book'. And yet Müller hoped that historians and general readers alike would pick up the book and read beyond its obvious faults of historical scholarship in order to see 'that there are in it whole chapters full of excellence, telling passages, happy delineations, shrewd remarks, powerful outbreaks of real eloquence, which could not possibly be consigned to oblivion'.[37]

This, of course, was the rub. What difference did it make if in managing to impart to his students something about the wonder and mystery of the past Kingsley did so through mythical parables rather than established fact? What difference did it make if in sharing some insight or some essential essence, some essential truth of the past, the particular facts were poorly interpreted or mishandled? The fact of the matter was that Kingsley's method was immensely popular with students, and while they may have failed to learn the basic principles of scientific history as conceived by someone like Freeman or Stubbs, they were excited by the study of history nonetheless.[38]

Kingsley was not entirely alone in his more artistic conception of history, one that seemed to privilege style and drama over substance and even accuracy. His close friend James Anthony Froude (1818–94) was often attacked for the same sorts of problems that seemed to plague Kingsley's historical writings. Froude, however, produced a mountain of historical studies, most notably his twelve-volume *History of England from the Death of Cardinal Wolsey to the Defeat of the Spanish Armada* (1856–70), in all likelihood the first history of England based on extensive archival research. One could be forgiven for thinking that Froude's historical methodology was diametrically opposed to that of Kingsley and more along the lines offered by Stubbs and the inductive set, particularly when reading Froude's methodological statements from his early historical work. 'It is not for the historian to balance advantages', argued Froude in the first volume of his *History of England*. 'His duty is with the facts.'[39] But an English Ranke Froude would not become.

Upon the first appearance of the first two volumes of the *History of England* in 1856, Froude was deemed an interloper just as Kingsley would be a few years later. Froude's name was quite well known by then as a controversial novelist who publicly attacked his respectable archdeacon father, the memory of his pious brother, as well as Oxford and the Church of England, all for the sake of celebrity and in order to sell some books. The reality of Froude's early life is much more complicated than was typically believed, but the controversy engendered by his first two novels followed him throughout his life and was often sparked anew by anything he wrote, whether on English history, on Carlyle, on the colonies, or on the art and science of history itself. Froude grew up in a deeply respected clerical family. His father was the well-known archdeacon Robert Hurrell Froude, a church administrator and strict disciplinarian. In his eyes, the youngest son, James Anthony could do absolutely nothing right, while the oldest son, Hurrell, could do little that was wrong. Hurrell tormented his much younger sibling (they were fifteen years apart) while the archdeacon found James Anthony to be weak in mind and spirit, particularly in comparison to Hurrell. Their mother, who had kept Hurrell in line as best she could, died of consumption when James Anthony was only three, thereby leaving the youngest son entirely at the mercy of a decidedly masculine household.

While James Anthony was being absolutely brutalized at public school, Hurrell was sent to Oxford and became close friends with Newman and Keble. Along with them, Hurrell became one of the early leaders of the Oxford Movement. When he died far too young of tuberculosis in 1836 Newman edited and published Hurrell's religious daybooks in four volumes as *Remains of the Reverend R. Hurrell Froude* (1838–9).[40] While the volumes were in Newman and Keble's view 'testimony of the first Tractarian saint',[41] they painstakingly recounted Hurrell's daily struggles to atone for nameless sins through fasts and self-flagellations that made him seem more like 'a Roman Catholic Monk (circa the Middle Ages)'.[42] The *Remains* proved to be highly divisive, forcing many evangelicals to repudiate any alliance they might have had with Tractarians in common support against the further liberalisation of the Anglican Church. The evangelical *Christian Observer* became suddenly hostile towards Tractarianism commenting upon the publication of Froude's *Remains* that 'The battle of the Reformation must be fought once more'.[43] 'Hurrell Froude's *Remains* were scandalous', argues Julia Markus. 'They called out, *Beware! The Papists are coming!*'[44]

Hurrell's father, however, was appalled by the criticisms. Despite the fact that he remained largely unconcerned with the specific theological debates engendered by his son and the Tractarians, the *Remains* were evidence of a deeply pious Christian cleric. The younger Froude could not live up to his brother's saintly image. Whenever he fell from grace, which was often, he was told to re-read his brother's *Remains*. James Anthony, for his part, much later claimed that the *Remains* was brought out by the leaders of the Tractarian movement as a mere 'party manifesto' rather than as a thoughtful memorial. The 'publication of the *Remains*', Froude claimed, was 'the greatest injury that was ever done to my brother's memory'.[45]

In attempting to live up to his brother's memory, however, Froude would follow his footsteps and attend Oxford just a few months after his death, even working with Newman. He was not a superb scholar but he did receive a promising second-class degree which was good enough for him to obtain a coveted fellowship at Exeter College; this required him to take deacon's orders, but also gave him time to write. He decided to write a supposed fictional account of his early life that would be published as *Shadows of the Clouds* in 1847. In order to avoid harming anyone portrayed badly in the book he published under a pseudonym, Zeta, and provided false names throughout. It was primarily about his brutal upbringing at the hands of his silent and disciplinary father as well as about the torture some children are forced to endure at public school. While the book did, by all counts, a brilliant job in exposing these very general problems in Victorian Britain, the book was 'so very obviously the life of Froude that there is no mistaking it', argued one of the future authors of *Essays and Reviews* Benjamin Jowett. The confessional and brutally honest tone made it practically a new edition of the late 'Froude's remains'.[46]

As a result Froude began to be alienated from his family. Oxford was also not happy about the controversy engendered by the book, though the dean at his college chose to accept Froude's claim that he was not the author. Two years later, Froude would publish *The Nemesis of Faith*, a book that Oxford could not this time ignore. It was of the same confessional and autobiographical genre as *Shadows* but the subject matter was largely religious, laying open for the public Froude's immense crisis of faith. The book was about a Church of England cleric just like himself who was not convinced by many Christian doctrines, including that of Christ's divinity. It was deeply confessional. Froude claimed of the book, sounding very much like Percy Bysshe Shelley, that 'I cut a hole in my heart and wrote it with the blood'.[47]

Nemesis was written as a series of letters by a confused young cleric, Markham Sutherland, to his trusted friend who is known only as Arthur. Just like Froude, Markham struggles with his decision to take deacon's orders claiming that he is unable to declare that he believes, according to the Thirty-Nine Articles, all 'the canonical writing of the Old Testament'. Markham was particularly confused by the wrathful God of the Old Testament who was 'not a Being to whom I could teach poor man to look up to out of his sufferings in love and hope'.[48] 'No, if I am to be a minister of religion, I must teach the poor people that they have a Father in heaven, not a tyrant; one who loves them *all* beyond power of heart to conceive; who is sorry when they do wrong, not angry'.[49] He also questioned the authority of the Bible itself and repeated many of the criticisms being voiced at the time by biblical scholars relying on new critical research methods. He argued that we must stop treating the Bible as if it did not have a history and stop 'making a very idol of the Bible, treating it as if we supposed that to read out of it and in it had mechanical virtue, like spells and charms-that it worked not as thought upon thought, but by some juggling process of talismanic materialism'.[50]

Markham's crisis of faith is unfortunately only worsened by confessing his concerns to Church authorities and family members. His dean and uncle in particular treats Markham's crisis as 'simply a disorder' that would best be treated by disciplining the mind through activity rather than with idle speculation. The message to Markham is clear: 'regular activity alone could keep soul or body from disease. To sit and think was simply fatal; a morbid sensitiveness crept over the feelings like the nervous tenderness of an unhealthy body, and unless I could rouse myself to exertion, there would be no end at all to the disorder of which I complained.'[51]

While this was the great age of crisis of faith literature, it was rare for a Church of England cleric to publish so openly about doubting the Christian faith. Not only that, but the protagonist is dismissive about the clerical profession. Instead of spending time thinking about the 'deepest and most absorbing interests of humanity', Markham lamented, clergymen have been cursed by turning their 'duty' of saving souls into a duty of merely making a living.[52] This was blatant criticism of Froude's chosen profession. While it may have been presented as a fiction, Froude did not this time hide behind the veil of a pseudonym, nor would he hide his current position in the Anglican Church. The book was signed J. A. Froude, M.A., Fellow of Exeter College, Oxford. The book caused a scandal and was publicly burned at Exeter College Hall. He was forced to resign his fellowship and his father stopped speaking to him. He was no longer welcome at family events.

For Froude, the writing of the 'book was an extraordinary relief. I had thrown off the weight under which I had been staggering. I was free, able to encounter the realities of life.'[53] He managed to come to terms with his religious doubts, rejecting the High Church Anglicanism of his brother and Newman as well as the Thirty-Nine Articles that were so central to the established Church. Having taken deacon's orders a few years earlier, however, meant that Froude was cut off from entering any other profession. And yet, having published *Nemesis of Faith*, he was unable to take up the lone profession he was legally limited to. This was clearly on Froude's mind while he was writing *Nemesis* as Markham wonders 'if I decline my living, what is to become of me?' He muses that perhaps he could write books: 'The men that write books, Carlyle says, are now the world's priests, the spiritual directors of mankind.'[54] Froude would have little choice but to embrace his talent as a writer, perhaps becoming one of Carlyle's literary priests in the process.

Stripped of his fellowship and alienated at home, Froude had few places to turn but it was Charles Kingsley who opened his arms to the disgraced cleric at great cost to his own reputation. Froude stayed with Kingsley for several months and eventually married the sister of Kingsley's wife. Kingsley's writing influenced Froude a great deal and the latter began reading Carlyle quite seriously, finding much to embrace in his hero worship and his poetic reconstructions of historical events and persons. It was during the early 1850s when Froude's writing shifted from the fictional to the historical as he found a more grounded genre for what many regarded as a brilliant writing style. One of the first essays he wrote was entitled 'England's Forgotten Worthies', a narrative that clearly showed the influences of both Kingsley and Carlyle as Froude set out to rescue the forgotten heroes of Elizabethan England from posterity. They were the great pirates of the era who Froude believed did much to preserve England's future. It announced, more importantly, a re-examination of the English Reformation. The swashbuckling seamen were not evidence of a decadent and degenerate age, as the Tractarians and other whig historians such as Stubbs and Freeman would have it, but heroes who sought to preserve England's right to determine its own future.[55]

This initial foray into the history of early modern England set Froude on a path to reconsider the English Reformation in its entirety, a project that would become his *History of England*. The first two volumes of that work set about reconsidering the reign of Henry VIII, a period of English history that had become in Froude's mind clouded by the many intrigues of the King's personal life to the point of distortion. Against the received wisdom of the Victorian era, Froude argued that Henry VIII was truly a great man, outmanoeuvring his many opponents both foreign and English, while establishing a national church as a way to preserve the autonomy of the nation against Catholic Europe and the overreaching tentacles of Rome. Froude had come a long way from his youth, when he was influenced by the Tractarian extremes of his brother and Newman. Yet controversy and his past continued to haunt him.

The first few volumes were absolutely demolished in a review by then Regius Professor of Modern History at Oxford, Goldwin Smith, in the pages of the *Edinburgh Review* in 1858. Smith believed that the book initially showed promise given that the author seemed to claim he was a member of the 'new school' of Rankean historians who sought to rid their work of 'brilliant rhetoric' in favour of 'minute accuracy of detail'.[56] Smith believed that Froude had a great advantage in fulfilling this method over previous historians of England given the newly available archival materials and cataloguing ventures that made accessible previously unknown or unverifiable facts of the past.[57] However, in Smith's view Froude did not live up to the scientific standard he set for his book. Smith suggested that Froude, instead of merely presenting the facts, strained his imagination throughout in his attempt to paint vivid pictures of past events. While Smith had to admit that Froude had 'a great command of beautiful imagery', and that there was a clear 'beauty of certain sentimental and poetical passages' throughout the work, he believed that Froude's style 'runs a little wild' in places, having the effect of forcing dramatic meaning where there should have been none. In this way, both in 'style' and in 'sentiment', Smith argued, 'Froude often palpably imitates Mr. Carlyle'.[58]

Still, there was a much graver defect at work in Froude's *History*, 'a pervading paradox of the most extravagant kind', one related to his overly stylized prose that Smith believed led to the most absurd of interpretations. Smith argued that Froude, having plumbed the newly available primary documents relating to his study, was simply not satisfied with adding the 'mass of new and interesting details' to the 'received historical view', nor was he willing to modify received views 'in regard to questions of a secondary kind'.[59] Instead, Froude wanted to make a 'great discovery...to reward adequately so much labour, and to satisfy the expectation raised by the opening of mines of documentary evidence hitherto unexplored'. This forced discovery, in Smith's mind, was Froude's argument that Henry VIII was a 'palimpsest', that the Tudor was, against all evidence to the contrary and even to common sense, 'a perfect king'. For Smith, Froude was trying to shock his readers, engender controversy and sell some books. Froude's argument also indicated that Froude had clearly succumbed to Carlyle's doctrine of hero worship. A more damaging critique could not be levelled at a historian claiming that his 'duty is to the facts'. Smith presented his review as if pulling the veil from Froude's rhetorical methodology, linking his exposed hero-worship to his earlier work as a novelist and to the muscular Christianity associated with Kingsley: 'By a most natural reaction the author of "The Nemesis of Faith" and "The Shadows of the Clouds" has now embraced "muscular Christianity," combined with "Hero Worship" of Mr. Carlyle, whose influence, as we have before mentioned, is visible in his reflections and style. Approaching the history of the English Reformation in this temper of mind, he could scarcely fail to be captivated by the strong will, the forcible language and the vigorous administration of the second Tudor.'[60] The message of Smith's review is that Froude was not guided by a proper research method, that he was instead directed by his own dramatic imagination, an imagination skewed by various forms of unbelief as well as muscular Christianity and the 'dangerous example' set by 'Mr. Carlyle'.[61]

Froude responded to Smith's review in the pages of *Frazer's Magazine*, refuting Smith's main criticism, that his interpretation of Henry VIII was imagined rather than found. He admitted that his conclusion concerning Henry VIII had 'been brought inevitably in collision with received opinions', but he was adamant that this was no simple 'interpretation' on his part, that it was rather 'the interpretation of contemporary statesmen' found in the State Papers. Those who would argue with such an interpretation 'must argue with better than a sneer: out of the material which the publication of the State Papers has placed within their reach, they must produce some fact or facts which will prove the Statutes to be untrue'.[62] Froude turned the tables on his reviewer, arguing that it was Smith who had a preconceived interpretation of Henry VIII. A good historian, argued

Froude, would not rely on received wisdom; a good historian would go to the sources and check for himself.

Despite Froude's persuasive response, the key points of Smith's review would continue to haunt Froude throughout the rest of his life. Froude could not adequately distance himself from his previous life as a controversial novelist, nor from the perception that he was an interloper merely attempting to sell books rather than find and disseminate the truth about the past. To many, he would always be that controversial boy writing autobiography. The fact that he relied extensively on archival sources was typically praised in a backhanded manner, but his use of such sources was always viewed with suspicion, as if his weak 'temper of mind' combined with his stylistic prose to transform found facts into created fiction. Professionalizing historians attempting to distinguish history from literature could not help but see Carlyle or, even worse, Kingsley when they read Froude.[63] But at least on the surface Froude appeared to promote the new inductive scientific method of fact-based research and objectivity even though, as Smith made clear, his work did not live up to his supposed methodological presuppositions. This would soon change, though it would not be the content of his work that did so.

In 1864, Froude gave a public lecture at the Royal Institution on pretty much the same subject matter Kingsley spoke on at Cambridge four years earlier, a lecture entitled 'The Science of History'. Unlike Kingsley, however, there would be nothing ambiguous or confusing about his opinions. Indeed, there was no wavering in determining whether or not history was, in fact, a science. He would use the subject matter as a foil to make clear both his methodological and epistemological opinions concerning history.

After apologizing for undertaking such a 'dry subject' Froude suggested that there was 'something incongruous in the very connection of such words as History and Science. It was as if we were to talk of the colour of sound, or the longitude of the rule-of-three.' Froude continued by criticizing who else but the 'eminent' Buckle 'whose name is connected with this way of looking at History'.[64] Despite the fact that Buckle 'had the art which belongs to men of genius', Froude believed that speaking of history as a science 'is an abuse of language'. Just because there is a science of other things does not mean that every form of knowledge must be considered scientific to be legitimate. Furthermore, Froude argued, the physical sciences require a repetition of phenomena in order to test hypotheses and establish laws, whereas for history this is impossible. History, Froude argued, could not become a science in the same sense that astronomy or physics were sciences.[65]

If Froude was dismissive of Buckle's history as a natural science, he was equally dismissive of the contemporary fashion to consider history as an empirical one. We cannot content ourselves with a mere scientific explanation of the facts of the past, argued Froude, because, for one thing, facts are not value-neutral nuggets that provide easy access to some long lost place. 'They come to us through the minds of those who recorded them, neither machines nor angels, but fallible creatures, with human passions and prejudices.'[66] The historian, then, cannot be expected to let the facts speak for themselves, given their already subjective nature and that of primary documents in which they are located.

Moreover, facts can also be used in a multitude of ways, argued Froude. They are not simply found by the historian, but are necessarily interpreted and even shaped to fit into a broader conception of the past, a conception that might be unspoken by the historian. 'You may maintain that the evolution of humanity has been an unbroken progress towards perfection', argued Froude, or 'you may maintain that there has been no progress at all, and that man remains the same poor creature that he ever was', but '[i]n all, or any of these views, history will stand your friend. History, in its passive irony, will make no objection. Like Jarno, in Goethe's novel, it will not condescend to argue with you, and will provide you with abundant illustrations of anything which you may wish to believe.'[67] Compare such remarks about the relative nature of historical inquiry with Froude's response to Goldwin Smith's review, where he argued that he was merely presenting what statesmen had written, that he had neither shaped nor interpreted the facts discovered in the archive. Here Froude appears to have abandoned his support of, or faith in, the empirical foundation of history's science in favour of an almost relativist conception of the historian's work.

Froude would back away from the more extreme interpretations of his argument, suggesting that there are lessons one can learn from the study of the past, whether it be a fiction agreed upon or not, and these are 'that the world is built somehow on moral foundations; that in the long run, it is well with the good; in the long run, it is ill with the wicked'. Froude was adamant, however, that 'this is no science; it is no more than the old doctrine taught long ago by the Hebrew prophets'.[68] Indeed, if the historian wanted to contribute to the great cause of the truth, argued Froude, it would be better to mimic the poet rather than the scientist. Historians must recognize the limitations of history, admit that it is an art rather than a science, and work to make it the best art it can be:

> Philosophies of history, sciences of history – all these, there will continue to be; the fashions of them will change, as our habits of thought will change; and each new philosopher will find his chief employment in showing that before him no one understood anything. But the drama of history is imperishable, and the lessons of it will be like what we learn from Homer or Shakespeare – lessons for which we have no words.[69]

This was a spirited defence of history's art against its science; the science of history, in Froude's mind, was merely a passing fashion of the time. By the end of the first decade of the twentieth century, Froude's critique of the science of history would have appeared prescient; but in 1864 it was no less than heresy.

Critics would agree that there was little ambiguity in Froude's argument, however wrong it may have been.[70] 'The propositions which [Froude] lays down will not be strange to any who have taken the trouble to observe the scope and purpose of his previous writings', argued the *London Review*. Froude, just like 'Mr. Carlyle [and] Mr. Kingsley, ... protests vehemently against the possibility of a science of history at all'. The reviewer appreciated the attacks on Buckle and was sympathetic to Froude's 'union of industry and imagination' but could not understand 'why an attempt should not be made to draw large conclusions from such useful inductions as our historians now supply with us'.[71] It was precisely this rejection of any conception of history as a science that particularly irritated historians unsympathetic to Froude's approach. For Froude, history was an art form, full stop. As it turned out, the Royal Institution lecture would confirm the nascent discipline's worst fears about Froude's historical intentions. To professionalizing historians, Froude appeared to be undermining much of the good inductive work being done even though – indeed because – his works were very popular with the general reading public. Making history a professional and scientific discipline of study seemed quite dependent on exposing Froude's work as itself a relic of the past, and over the next thirty years it became the negative example of – and the foil for – proper scientific history.

4 DISCIPLINE AND DISEASE; OR, THE BOUNDARY WORK OF SCIENTIFIC HISTORY

... history only becomes interesting to the general public by being corrupted, by being adulterated with sweet, unwholesome stuff to please the popular palate.

J. R. Seeley, 'History and Politics IV', *Macmillan's Magazine*, 41 (1879)

I am fully convinced that it is not conscious lying ... but a strange mental and moral twist.

Edward A. Freeman on J. A. Froude, 27 April 1879 (letter to J. R. Green)

Making history a science not only involved the expression of a method that historians could ascribe to at will. For history to become a true science, the scientific method had to become a communal method – a *mos communis* – that would in essence form the very basis of a professional discipline of history. Making the inductive scientific method *the* method of history would involve a fair amount of disciplinary acts in the form of epistemic boundary maintenance. In the late 1850s and early 1860s, the boundary work of scientific historians seemed to focus on separating their inductive endeavour from Buckle's more philosophical and deductive project. As we have seen, however, historians promoting history's more artistic sensibilities also criticized Buckle's science of history as a way of rejecting history's scientific status entirely. What is more, such historians tended to promote a way of writing and teaching history that was much more popular with students and the reading public than was the fact-based approach advocated by the likes of Stubbs and others. It is for this reason that during the late 1860s through to the 1880s and beyond, the boundary work of scientific historians shifted towards demarcating history as a science from history as a form of art. A proper inductive analysis of the past was serious business, according to the Rankeans, not the trifling entertainment sought by the reader who picks up the work of Carlyle, Macaulay, or one of their many imitators (e.g. Froude). The goal of the Rankean-inspired scientific history was not to entertain the public with a fascinating story; its goal was to present the past as it actually happened, and this would necessarily make proper historical analysis a subject matter likely less interesting to the general reader than the shoddier, more dramatic work of Romantic historians. Convincing the wider public as well as other historians about the utility of scientific history despite its less immediate appeal than the more literary and artistic approaches was a central strategy in the writings of a few key proponents of a Rankean science of history during this period. Exposing Romantic history as the work of half-wits was an important element of this strategy.

History's primary disciplinarians during this period were J. R. Seeley and Edward A. Freeman, ironically two of the most popular historians of the second half of the nineteenth century. Despite the fact that both Seeley and Freeman failed to live up to the scientific standards they themselves promoted, they established a normative identity for the scientific historian through their punishing criticisms particularly directed at those who promoted an altogether different identity: the historian as artist.

The first real casualty of the boundary work of scientific history was Charles Kingsley. Despite the fact that his lectures were packed with excited undergraduate students, he was viciously attacked in the periodical press as an interloper with absolutely no business teaching history to the nation's youth. The attacks on Kingsley began at the moment he was appointed but they became particularly urgent once Stubbs became Regius Professor of Modern History at Oxford in 1866. For Rankean historians, it appeared that Oxford finally had an individual devoted to the specialized study of historical documents in a symbolic position of authority while history at Cambridge was being taught by a controversial novelist devoted to making history a form of art, a mere literary endeavour. Kingsley finally resigned in 1869 citing ill health and the strain of lecture writing, though it is also apparent that he could no longer endure Freeman's continuous attacks.[1]

John Robert Seeley (1834–95) replaced Kingsley as Regius Professor of Modern History at Cambridge in 1869 though this did not appear to be the great victory for scientific history that it would in hindsight become. This is because Seeley did not appear to be much more qualified than was Kingsley in teaching to the Cambridge syllabus. What was perhaps worse was that Kingsley had handpicked Seeley to be his replacement. The rationale for Kingsley's endorsement was also extremely concerning especially for High Church critics. It was less Seeley's historical work that impressed Kingsley and more the theological views that Seeley published in his anonymous *Ecce Homo* (1865), an enormously popular work on the life of Jesus Christ that he would ultimately regret publishing. The book itself caused much controversy, not least because of its mysterious authorship, and by 1869 Seeley had yet to overcome it, though he eventually would.

Seeley's religious beliefs overlapped quite closely with those of Kingsley. He was a liberal Anglican and like Kingsley he embraced scientific advances in all subject matters and yet felt that Christianity could only be strengthened by such scientific work. Christianity and science were complementary, according to Seeley, in the sense that they had the same goal: achieving self knowledge for the purpose of character reform.[2] He was heavily influenced by biblical criticism and David Strauss's *Das Leben Jesu* (1834) but his own history of Jesus Christ was less a critical examination of biblical documents than an analysis of Christ the man, the moral philosopher. Seeley set out to ignore the theological speculation about Christ, the miracles and Christ's divinity, and simply wrote about what Christ thought and how he acted. He argued that Christ's moral teachings should provide the foundation for a new science of politics, and even the foundation for a true Christian state governed by a universal positive morality that would also embrace the 'blessed light of science, a light ... dispersing everyday some noxious superstition, some cowardice of the human spirit'.[3] This turned out to be a divisive message, one that found outspoken detractors and supporters alike.

Critics could not understand how Seeley could separate the moral philosophy of Christ from the miracles. Some found this logically inconsistent;[4] others found it verging on the heretical.[5] Seeley knew his book would be hated by many of both the High and Low Church and he wanted to protect his family's name from the probable backlash. He was also fairly certain that many of his devotedly evangelical family members would have found his portrayal of Christ heretical. His father Robert Benton Seeley was an active philanthropic evangelical who staunchly defended the Church of England in his book *Essays on the Church* which had reached its seventh edition by 1859. He was, like many evangelicals of his kind, violently opposed to Roman Catholicism on one hand and just as viciously opposed to anything that had an odour of liberalism to it, whether of theology or politics. That *Ecce Homo* seemed to be just the mixture of liberal Anglicanism and Christian socialism promoted by the likes of Kingsley, Seeley knew his family would be embarrassed and ashamed to have the Seeley name associated with it.[6]

Many readers, however, were very appreciative of what they believed to be Seeley's deeply Christian message. Seeley received quite a bit of fan mail, largely from Christians who believed his presentation of Christ accorded with their own views.[7] The Congregational Minister, John Christien, wrote that '[a] mingled feeling of profound thankfulness and admiration has been awakened' by his reading of *Ecce Homo*.[8] Kingsley was one of the book's most outspoken supporters and he had his whole family reading it. His daughter Rose wrote to publisher Alexander Macmillan expressing her love of the book and her curiosity as to Seeley's identity, while Kingsley's wife went so far as to suggest that *Ecce Homo* was 'the most important book' that she had 'ever read except the bible'.[9] Macmillan also received a supportive letter from the future Prime Minister W. E. Gladstone, who would write some of the most glowing reviews of the book, reviews that would later be published in book form.[10]

Even though Seeley received much support, it was difficult for him to endure the criticisms as well as the endless questions as to his identity. He did not seem to care that the book was an unexpected success, going through an astonishing six editions in the first full year of its existence. In terms of the book's sensation, Gladstone put it in good company:

> No anonymous book, since the 'Vestiges of Creation,' (now more than twenty years old) – indeed, it might almost be said no theological book, whether anonymous or of certified authorship – that has appeared within the same time interval, has attracted anything like the amount of notice and of criticism which have been bestowed upon the remarkable volume entitled 'Ecce Homo.'[11]

But this was attention Seeley wanted to avoid.

The surprising success of the book seemed to catch both Seeley and his publisher off guard. They clearly had not developed a strategy for dealing with questions about the author's identity and in this regard Macmillan was often more interested in selling the book than in protecting his author.[12] Eventually too many individuals were let in on the secret and Seeley's authorship became widely known. In November of 1866 the *London Review* claimed the authorship must be 'Professor Seeley, of University College, London'.[13] By December of 1866, barely a year after *Ecce Homo* was first published, Macmillan wrote to Seeley that the secret was clearly out and felt there was little reason to keep up the charade: 'I suppose I need make no mystery now, need I?'[14] Macmillan, it seems, believed that the secret would come out sooner or later and that Seeley would eventually want to sign his name to a future edition. Seeley, however, refused to do so, even after his authorship became common knowledge. By this time he had become very frustrated by much of the negative press, and was simply dumbfounded at the ridiculous misrepresentations of the book's meaning perpetuated by many readers and reviewers. 'The success of Ecce Homo was rather alarming than otherwise', Seeley later wrote to Macmillan. 'If I knew any way in which [to] prevent all weak or rash heads from reading it, I would certainly adopt it.'[15] This concern of excluding 'weak or rash heads from reading' his work would become a much more prominent theme in Seeley's life as a historian.

By not openly attaching his name to *Ecce Homo*, Seeley was able to avoid overt public persecution but this certainly did not preclude him from benefiting from the few powerful friends who did happen to like the message of *Ecce Homo*, and who were convinced of his authorship. Not only did Kingsley tap him on the shoulder as his successor as Regius Professor but in 1869 Gladstone was Prime Minister and it was his ultimate decision that led to Seeley's appointment.

Unfortunately for Seeley, it was all too clear that he was offered the position largely on the basis of an anonymous publication that was not really a work of historical scholarship. Seeley's appointment, much like Kingsley's nine years before, was controversial.

The *Saturday Review* felt that Seeley's appointment was further evidence of a disturbing trend at Cambridge, a university that simply did not take the study of history seriously. Seeley's appointment, however, was also troublesome because of Gladstone's well-documented approval of *Ecce Homo*, the anonymous work that seemed to secure Seeley's appointment – a patronage appointment if there ever was one. 'Mr. Gladstone was fascinated with Ecce Homo, and therefore Mr. Seeley teaches modern history at Cambridge. He may do it well; but his nomination was quite independent of any sufficiently grounded presumption that he would do so.'[16] The *Saturday* went on to explain how previous office holders at least had a publishing record, whether literary or historical, that placed them in an elite class of writers. Seeley, however, 'is not one of this order. He owes his appointment to the personal liking and to the theological sympathy of the Minister of the day. He has still to prove his fitness for the place he fills.'[17]

By most accounts, his inaugural lecture did not inspire much hope that the appointment was somehow a fluke, though the *Athenaeum* approved.[18] Seeley spoke on a theme that would dominate much of his historical thinking for the rest of his life: the relationship between history and politics. In an argument that for the *Saturday* reeked of 'vulgar utilitarianism',[19] Seeley argued that history had yet to become a science because it was largely viewed as mere entertainment and not valued for what it can teach about political matters. History, properly understood, should be the training ground for statesmanship. The *Saturday* found this argument 'too precarious to form the basis of an argument as to the uses of the study of modern history. ... Modern history is, or ought to be, something better than the handmaid of contemporary politics.'[20] J. R. Green similarly complained that a training in history should be first and foremost in preparation of becoming a historian, not a politician. Even 'Kingsley never talked such rubbish as this'.[21] The Master of Trinity, Montagu Butler, also found himself surprised that 'we should so soon have been regretting poor Kingsley', a quip that G. R. Elton called 'the finest double anti-compliment ever uttered'.[22]

When Seeley's early Cambridge lectures were published in late 1870 opinions about his historical capabilities began to change. The *Saturday* revised its position on Seeley, albeit only slightly, arguing sarcastically that 'Seeley, like many other people, is better than his own doctrines, and that he himself has not wholly neglected the studies on which he teaches the students of Cambridge to set so little store'. However, and perhaps in deference to Stubbs, the *Saturday* found that the volume did not give 'the impression of really hard work. We seem to be listening to a man who has rather played with his subject than toiled at it.'

The lectures did indicate, though, that Seeley did have some promise and while the *Saturday* was not quite ready 'to admit Mr. Seeley as a full guild-brother of the order of historical scholars', it was ready to 'gladly welcome him as an outsider who keeps his eyes open, and the result of whose speculations we are thoroughly glad to hear'.[23]

Throughout the 1870s, Seeley's stock within the professionalizing historical community would continue to rise. While Kingsley was generally uninterested in the history curriculum, Seeley was profoundly concerned with the way in which history was taught throughout the colleges at the undergraduate level. Not only was he central in building a Historical Tripos at Cambridge in 1873 that effectively established history as an autonomous discipline of study there, he ensured that the curriculum combined specialized topics with general subjects related to history such as political economy and political philosophy. Indeed, it was not long before history teaching at Cambridge reflected the past as politics vision of its professor of modern history.[24] Freeman certainly approved.[25]

Seeley gained further acceptance with the publication of his dense three-volume *Life and Times of Stein, or, Germany and Prussia in the Napoleonic Age* in 1878. The work was largely derivative and yet the *Saturday Review* found it to be 'a valuable contribution to English knowledge of German history and German politics'. The *Saturday* was particularly impressed by the clear 'patient industry' necessary to complete such an 'exhaustive study' with such 'discriminating impartiality'.[26] This was high praise from a journal that only eight years earlier found his work lacking the appropriate amount of toil. It certainly helped matters that Seeley was at this time writing quite critical essays that challenged the notion that history was a literary endeavour while promoting a much more specialized discipline of history based on inductive scientific methods.

In 1879 Seeley wrote a series of articles for his publisher's liberal house magazine, *Macmillan's Magazine*, entitled 'History and Politics' that were meant to 'present some of the more general views about the study and teaching of history' by a 'specialist' who has 'been engaged for ten years in teaching history at one of our great universities'. He wanted to direct his message not to the 'general reader nor to the pure scientific theorist, but rather to those engaged in the higher education – those who inquire practically what place history is to fill in our national culture, and how the teaching of it ... may be made more reasonable and more useful'. In order to do that Seeley claimed that it was necessary to discuss the 'two broad movements' that 'are now observable in the historical world': the movement to make history 'accessible and readable'; and the movement to give history the 'exactness of a science'. Seeley was quite up front that he believed historians needed to work on achieving the latter at the expense of the former.[27]

Seeley argued that there is a general misconception that making history a science will necessarily turn history into the kind of discipline Buckle envisioned, one that focuses on the masses and relies on statistical regularities to develop laws to explain the historical process. This is an older scientific view, argued Seeley, one that relies far more on speculation than scientific methods. He admitted that there might be a place for the kind of work Buckle promoted but such would not be history.[28] Seeley's science of history simply 'collects and verifies facts in order to draw conclusions from them'. But, perhaps more importantly, '[i]ts aim is not to give pleasure or confer fame, but to throw light on the course of human affairs'.[29] This, of course, is the exact opposite of the aim of a history explicitly written to titillate a popular audience, the real target of Seeley's polemic.

Seeley went on to explain that if Buckle's work managed to popularize a false conception of the science of history, the work of both Walter Scott and Macaulay popularized a false conception of history itself. Seeley had to admit, however, that he appreciated Scott's historical novels because Scott 'brought history home to people who would never have looked into the ponderous volumes of professed historians'. Indeed, he believed that there were quite simply 'large historical periods which would be utterly unknown to us but for some story either of the great romancer or one of his innumerable imitators'. But Scott was almost too successful in making the past come alive influencing perhaps too much of both the writing and reading of history.

> Writers, as well as readers, of history were awakened by Scott to what seemed to them the new discovery that the great personages of history were after all men and women of flesh and blood like ourselves. Hence in all later historical literature there is visible the effort to make history more personal, more dramatic than it had been before. We can hardly read the interesting Life of Lord Macaulay without perceiving that the most popular historical work in modern times owes its origin in a great measure to the Waverley Novels.[30]

Seeley argued that Macaulay was so influenced by Scott that he would compose 'imaginary conversations among historical persons ... like those in the Waverley Novels' in an attempt to 'make history as interesting as romance'. 'In a bookish age', Seeley argued, where novelists were praised for their 'magic skill' in making 'fiction look like truth' a 'magician' named Macaulay appeared and reversed 'this feat' and 'charm[ed] mankind equally by making truth look like fiction'.[31]

If Scott is to be praised for founding the 'historical romance' he is to be denounced for founding 'the romantic history'. Seeley went so far as to claim that Scott, and Macaulay after him, were responsible for engendering the popular misconception 'that the best book is that which is most readable. It is inconceivable to the popular mind that a man should write a book which it is difficult to read, when he might have read a delightful and fascinating one.'[32] Seeley pointed out that most good history is 'ordinary' and 'monotonous', 'ruled by routine', 'acts of parliament, budgets and taxation, currency, labyrinthe details of legislation and administration; topics, in short, which become the most tiresome in the world as soon as they have passed from the order of the day'.³³ Seeley explained that the 'serious student of history' has to realize that real history is simply not that exciting and 'submit to a disenchantment like that which the experience of life brings to the imaginative youth. As life is not much of a romance, so history when it is studied in original documents looks very unlike the conventional representation of it which historians have accustomed us to.'³⁴

Seeley argued that Romantic historians such as Macaulay made it extremely difficult for good historical work to be recognized as such. Macaulay's 'splendid success' in 'making history interesting has done a mischief which it is now very difficult to repair. It has spoiled the public taste, and in the natural course this corruption has reacted upon the writers of history.' Readers expect to be entertained as if sitting in a 'theatre ... gaz[ing] at the splendid scenery and costume' without the slightest regard for accuracy or truth. For Seeley this was problematic because history could not be perceived as an autonomous academic discipline. 'The final result is that to the general public no distinction remains between history and fiction. ... History in short is deprived of any, even the most distant association with science, and takes up its place definitively as a department of *belles lettres*.'³⁵

In his examination of Romantic history, Seeley slightly explored the style of writing that is employed but only to denounce it as necessarily engendering falsehood. The 'natural effect' of applying poetry to 'historical facts ... is to transform them into fables'. Seeley actually went on to examine the few simple rules that the historical poet must follow in order to make 'history interesting and exciting': 'All that is necessary is systematic exaggeration and occasional falsification.'³⁶ For Seeley, any form of writing that made history more readable or more interesting could only falsify. He did not argue with the fact that 'Romantic or readable histories ... diffuse a certain knowledge of historical names characters, and scenes', but he could not support the notion 'that they convey solid instruction. Nay, what is instructive in history is precisely that which it is difficult to read, that which cannot be understood without an effort, and this is what the readable history omits.'³⁷

Seeley seemed to suggest that it would be impossible for the general reader to understand history properly and therefore any attempt to reach such a reader should be avoided. '[A]s I look upon history as a scientific subject', Seeley explained, 'I do not hope that the general public can ever conceive it rightly'. He went one step further, arguing that '[a]ll direct attempts to popularize historical knowledge seem to me likely to fail, for history only becomes interesting to the general public by being corrupted, by being adulterated with sweet, unwholesome stuff to please the popular palate'.³⁸ His disciplinary counter-part, Edward A. Freeman, thought that perhaps Seeley went a bit too far here, that in attempting to do what he could not with *Ecce Homo*, that is to keep 'weak or rash heads from reading' his work, 'we ought to make history so dull and unattractive that the general public will not wish to meddle with it'.[39] For Freeman, history need not be boring but it must be above all accurate, and here he and Seeley certainly agreed. '[D]are to be accurate' was Freeman's first precept in engaging in real historical work, his ecclesiological pedantry converted into a decidedly Rankean principle.[40]

There was no question in Freeman's mind about history's scientific status. He argued that as long as we do not follow the rather strange contemporary way of confining the term 'science' to 'certain branches of knowledge' it is only too apparent that history is a science. He advocated thinking about science in much broader terms, as *scientia* or knowledge, rather than the current fashion of submerging 'science' with 'natural science'. He very much would have agreed with Acton in considering history a form of *Wissenschaft*. 'We have too deep a regard alike for the English and the Latin tongue to wish to be called *scientists*', argued Freeman, 'but we do claim for our studies a place among the sciences'.[41] Freeman believed that it was necessary to understand that history was not a science in the same way that, for instance, astronomy was a science. Historical evidence simply does not allow the historian to reach the same kind of certainty. And yet, Freeman argued, history as 'a science depends on the facts of history' despite the fact that they 'are not so nearly certain as the facts of some other branches of knowledge'.[42] The science of history, then, was highly dependent for Freeman on the way in which the facts themselves are handled and interpreted. It irritated him to no end the way in which Romantic historians too easily mishandled the facts for the sake of dramatic narrative thereby undermining the historians' claims to producing scientific knowledge.

It was for this reason that Freeman often lamented the necessary narrative form of historical writing. He famously wished that his readers could avoid reading his prose altogether by going straight to the authorities. This, of course, was impossible. Instead Freeman believed that it was simply necessary for the historian to acknowledge the unfortunate but 'unavoidable connexion between history and literary style' and to always be on guard for the 'temptations' that go along with it. 'There are temptations which beset the writer himself and temptations which beset readers.' The historian, Freeman argued, must fight above all against the temptation to make his narrative more enjoyable in order to maintain accuracy above all else:

> There is the constant danger that [the historian] himself may sacrifice accuracy to effect, that he may exaggerate something, that he may leave out something, that he may throw in some epithet or give his sentence some turn ... For as soon as any composition assumes a literary form, such readers will, or at least may, follow; and the temptation arises to give them, not the food which may be best for them, but the food which will most please their palates.[43]

'We must always be on our guard', Freeman argued, 'of that literary character of our work which cannot be avoided'.[44] For Freeman, Romantic writing could only lead to '*pseudo*-historical statements' whereby 'the narrator is either himself deceived or he intentionally seeks to deceive others; in purely romantic statements deception hardly comes in any other way'. This not only means that the teller of the story is 'simply careless about historical truth.' Freeman implicates the 'hearer' in this as well.[45] In defence of Samuel Gardiner's *History of England from the Accession of James I to the Outbreak of the Civil War* (1863–87), a work criticized as extremely boring,[46] Freeman suggested that the problem was not with Gardiner's writing style but with the critics more generally: 'Those who call him *dull*, I suppose want him to rave like Carlyle, or talk namby-pamby like Froude. That he won't, or I either.'[47] For Freeman, Gardiner's style of writing was first and foremost trustworthy: 'he impresses me with the feeling that he is telling a true story'.[48]

Nothing irritated Freeman more than historians who would give into the temptation to make history more interesting than it already is thereby sacrificing truth to effect. Despite the fact that he was one of the first English historian to devote his entire life to the study of history, he was forced to watch many lesser historians, in his mind interlopers, be granted the fast track while he was often seen as a mere pedant and likely unable to excite students to the study of the past. Indeed, Freeman's correspondences with his friends, colleagues and publishers are filled with references to the few university positions that were available for historians throughout his adult life. He even constructed ideal scenarios whereby he and his friends would be appointed to all of the top historical positions in the nation.[49] While his focus on achieving such a post might seem to border on the obsessive, he simply had no interest in doing anything else, and he was determined to make it happen.

One of the first positions to which Freeman applied was the Chichele Professor of Modern History at Oxford in 1862, a position created in response to the Oxford University Commission of 1852, which proposed that new professorships be founded with the assumption that those appointed would keep the position for life and devote their time to research and writing. It was thought that such professors would teach much less than the tutors who spent the majority of their time preparing students for examinations, with little time for their own research and writing. Professors, rather, would engage in furthering the knowledge of the particular discipline. The Chichele professorship was a product of the Commission. The professorship, however, was awarded to Montagu Burrows, an individual clearly not up to the standards Freeman himself set for historians and it was a brutal reminder about the clear gap between history teaching at the universities and the imagined historical community constructed by Freeman. As far as Freeman was concerned, Burrows was merely a 'tutor' and a 'coach' and not up to the standards of many of the other historians who had put their names forward for the position.⁵⁰

Those other apparently more qualified applicants included Freeman himself, Stubbs and even the much maligned Froude. The appointment of either of those candidates would have likely appeased Freeman and his historical friends (Froude had yet to become an enemy), but they could not garner the necessary support at Oxford. Burrows was backed by Bishop Samuel Wilberforce of Oxford as well as Prime Minister Gladstone who was ultimately responsible for the appointment. Burrows was considered an excellent tutor and despite the fact that the new position did not entail any tutorial functions it was his outstanding teaching that, at least partly, led to his appointment. The other factor in his favour was a book he wrote entitled *Pass and Class* (1860), which was essentially a practical manual to help students do well on exams.⁵¹ It is clear that the electors were looking for a good teacher rather than a good historian. Or, rather, being a good historian meant that one had to be a good teacher. This is most clear in their rationale for turning down Freeman's application.

In his letter to the electors, Freeman said almost nothing about his work as an examiner at Oxford; he concentrated on what he considered his past (and future) accomplishments as a professional historian. 'Since 1845, if indeed I might not say from an earlier time still, I have made historical study the main business of my life. I have both studied history as a whole, and I have more minutely studied several periods in detail, relying on original authorities alone.' Freeman was attempting to establish himself as a Rankean historian. At one point he mentioned that '[i]n the years 1857–8 I filled the office of Examiner in Law and Modern History in the University' but he did not make any other references to his tutorial record. He went on to try and explain the fact that he had yet to write a great historical work, a mark of a good historian by the new Rankean standards: 'If I cannot rest my claims on any great published work it is because I have always felt that I should be wanting to my subject if I ventured on such a work before I felt fully conscious of the maturity of my powers and information.' But, Freeman argued, he felt that he had now reached a certain level of historical maturity and could engage in that study, 'the history of Federal Government from the earliest times to our own day – which I trust will not discredit the ten years' study of which its first portion will be the result'. The rest of the letter highlights some of Freeman's published lectures and essays as well as his claimed interest in seeking the position which was not for the income or even 'as a means of professional advancement' but simply to do 'some service to the study to which I have devoted my life'.⁵² Freeman presented himself as a disinterested historical observer.

Freeman did not quite grasp the audience he was writing to. The electors were not looking for an historian to devote his life to the minute study of primary documents of a special subject or to someone who had published a great deal (as the rejection of Froude would suggest). The electors were looking for a good teacher and as the dean of University College (where Burrows taught) suggested, Burrows would be an excellent choice because 'a good modern history professor must ... in very great measure unite the tutorial with the professorial functions' something Burrows would do well.[53] Freeman mistakenly assumed that the electors were looking for a new Rankean historian to fill the position when, in fact, the electors did not know what to make of Freeman's letter or the professional identity he tried to construct therein.

Burrows's appointment was not well received by Freeman and his friends because this was one of the first positions available, or so it was assumed, for the professional historian. The *Saturday Review* could not understand how such an error could have been made. 'A fortnight has elapsed since the appointment of the Chichele Professor of Modern History, but the astonishment and grief which it has produced among the friends of Oxford and of learning have scarcely abated.' The 'friends of Oxford' who were so astonished and grief stricken is meant to refer to the historical community at large, but more particularly, the candidates not chosen. The appointment could have been justified, the article went on, 'had there been no eminent men among the candidates for such a chair'. But such was not the case:

> At least one distinguished historian [Froude], and at least one brilliant essayist [Freeman], were glad to offer themselves for the honourable position enjoyed by an Oxford professor; and there was at least one student [Stubbs] who might justly have looked upon one of the most lucrative chairs in the University as a fair reward for the devotion of many years to the successful study of the monuments of early history.

While Froude was no friend of the *Saturday Review* his appointment could have been justified. As laborious as it must have been to write, the *Saturday Review* had to admit though not without backhanded jibes that 'however false may be the canons by which [Froude] permits his historical judgment to be guided, [he] is at least master of a beautiful style, an indefatigable student and a man qualified even by his errors to give an impulse to the study of history'. Instead of hiring Froude, Freeman or Stubbs the electors chose a man 'who had not only given no proof of his fitness for the post, but had given the most satisfactory proof of the contrary'.[54]

The problem with Burrows, the *Saturday Review* claimed, had nothing to do with his tutorial work; in fact, the *Saturday* had to admit that 'Mr. Burrows is able to render really valuable assistance to his pupils'. The problem was that Burrows could not be considered a history professor but merely a 'popular' tutor 'without the pretence of possessing either special knowledge or the capacity of acquiring it'. The *Saturday Review* claimed that Burrows's *Pass and Class* was symbolic of his position as a tutor rather than professor and should have been the rationale for *not* hiring him.

> 'Pass and Class', the only book the new Professor has yet given to the world, is a little handbook of advice to people who wish to 'get firsts'. It combines with the sort of advice as to the choice of books which a tutor would give his pupil in the first hour of their intercourse, a number of very feeble remarks on the books themselves, and the suggestion of some rather ignoble tricks to be used in reading them. There is, indeed, nothing in 'Pass and Class' to show that Mr. Burrows is unlikely to be an efficient 'coach' for a school in which a high standard has not yet been reached, and it is said that the new Professor's pupils have been successful in the school of modern history. But there is ample evidence on every page of the book that its writer is not a man either to study history judiciously himself, or to create an enthusiasm for it in others.

The *Saturday Review* doubted that the electors had even opened the book. Of course, the electors did read *Pass and Class*; it was one of the main reasons for Burrows's appointment which suggests just how diametrically opposed the *Saturday Review* and the electors were in understanding not just the requirements for such a position, but the nature of professional history in general. The *Saturday Review* expected appointed historians to professional chairs to be the equivalent of chairs 'that in Germany would have been occupied by a Niebuhr and a Mommsen' rather than the 'coaches' Oxford had appointed.[55]

Despite the negative reaction to Burrows's appointment, his lectures were well attended. In his autobiography, Burrows noted that his 'inferiority to the historians' he had beaten out for the appointment was no doubt apparent to 'hostile critics' because their 'reputation was fast growing'. However, the electors were still correct in their decision because 'it was notorious that neither Stubbs nor Freeman, who successively became Regius Professors, could ever keep a class together … I at least kept an average attendance of 20 men during many years, and published books or articles in leading reviews every year.'[56] The electors wanted a successful teacher and that is exactly what they got by appointing Burrows.

The Burrows appointment, and no doubt the Kingsley appointment two years before it at Cambridge, hardened Freeman to the fact that obvious merit was simply not enough to achieve a coveted chair of history. He and his imagined historical community needed to put forward their new Rankean standards more explicitly and even violently. They needed to engage in the process of boundary work and expose the historical impostors while promoting the right and proper historical research methods. Kingsley made a fairly easy target for this and we have seen how Seeley used Macaulay's work as a foil to promote his particular brand of scientific history. Freeman, however, seemed to focus his attention obsessively on Froude's work, exposing the many errors of detail necessarily engendered by privileging style over accuracy. Froude became the primary target for Freeman's boundary work.

Boundary work, according to Thomas Gieryn, can function in a variety of ways depending on the particular professional goal in question. If the goal is to *expand* authority into previously unoccupied terrain, then boundary work 'heightens the contrast between rivals in ways flattering to the ideologists' side.' If the goal is *protection of autonomy*, boundary work functions to exempt 'members from responsibility for consequences of their work by putting the blame on scapegoats from outside'. Finally, if the goal is '*monopolization* of professional authority and resources, boundary-work excludes rivals from within by defining them as outsiders with labels such as "pseudo-," "deviant," or "amateur"'.[57] It is this last function of boundary work that is most applicable here. Gieryn's example in this instance is of the mobilization of anatomists in early nineteenth-century Britain against the burgeoning 'pseudo-' science of phrenology. Gieryn shows that phrenology was attacked for three main reasons: because it challenged certain orthodox theories and methods at the heart of contemporary anatomical practice; because its chief practitioner, George Combe, believed that truth should be certified by popular opinion, thereby undermining concepts of scientific authority; and, finally, because it sought to meld science and religion in a way that would have granted greater authority to religious conceptions of nature. In consequence of these threats, Combe was denounced as an interloper in much the same way as was the anonymous author of *Vestiges* and phrenology itself was deemed a pseudo-science. In this case, the boundary work was successful as Combe was denied the chair of Logic at Edinburgh University, which he greatly coveted, and phrenology was excluded from anatomical science's epistemological boundaries.[58]

In the context of a professionalizing history, Froude represented a similar threat to history as Combe did for anatomy. Froude believed that the scientific ideals of professional history were unattainable, that the true method of history was that of the artist. This not only undermined the scientific rhetoric of professionalizing historians, it also undermined their attempt to separate history from other literary genres. Furthermore, Froude's promotion of a dramatic style of history, combined with the immense popularity of his work, suggested that he was placing far too much power in the hands of readers and publishers in determining historical truth. Rather than argue with Froude – that his artistic methodology was not as fruitful as the new scientific one – historians set out to prove instead that Froude was not only a bad historian, but also that his many faults excluded him from the historical discipline entirely. Denouncing Froude as a pseudo-historian proved much easier than actually engaging with and criticizing his alternative method. In the long run, however, historians would not be as successful in excluding Froude as anatomists were in excluding Combe, though the effort was certainly there.

While Froude's supposed scientific method had been exposed by Goldwin Smith in the Whig quarterly *Edinburgh Review* (see Chapter 3), the campaign to expose Froude as an interloper really began in the pages of the *Saturday Review*. This was particularly the case when Freeman took over the reviewing duties of the *History* when Froude's volumes on the reign of Elizabeth appeared in 1863. If Froude was angered by the tone of Smith's review six years earlier, he must have been livid at the flippant and patronizing prose of Freeman. Freeman began his review by 'laugh[ing] at the ludicrous misapplications of evidence' that led to Froude's theory of Henry's 'infallibility'. Freeman admitted, however, that he was at least impressed that Froude took a side even if his argument was only made to attract a large audience more interested in drama than facts, pleasing 'those who take a pleasure in pretty talk about streams and blasts and daisies and dark November days and that mysterious clock which was always on the point of striking and yet never did'. Freeman went on to cite Smith's review, explaining that Froude's ridiculous thesis 'was dashed in pieces in the pages of the *Edinburgh Review* by a hand which evidently knew alike when to smite and how to smite'. Unfortunately, 'Froude and his admirers', argued Freeman, 'did not know that they were smitten' and the volumes continued unabated.[59] Freeman felt it his duty to continue the smiting.

Freeman admitted to being impressed by the extensive archival research that had gone into Froude's work, but he believed that Froude was unable to understand most of it because of his lack of proper training in the new critical research methods. This lack of a 'proper apprenticeship in historical writing' was unfortunately compounded by what Freeman referred to as Froude's 'natural' and 'incurable defects'. 'He lacks that calm and judicial intellect', argued Freeman, 'that love of truth at all hazards ... He has not the stuff in him that could ever guide him to ... unfailing accuracy and unswerving judgment.'[60] Froude, in other words, was inherently incapable of being or even becoming a historian, his natural defects necessarily excluding him from Freeman's imagined historical community.

The boundary work against Froude reached its climax in a series of review articles written by Freeman for the independent-minded bi-monthly *Contemporary Review*, a ninety-nine page slaughter of Froude's 'The Life and Times of Thomas Becket', a series of articles that appeared in the *Nineteenth Century*.[61] This set of reviews was typical of previous anti-Froude diatribes: it argued that Froude was a newcomer to history-writing (even though he had been publishing in the industry for more than twenty years); that, despite his archival work, he was an amateur when it came to research; that he was untrustworthy with the evidence; that his stylistic prose often over-dramatized what actually happened;

and, finally, that he was consistently inaccurate throughout his portrayal of the life of Becket. The nastiness of this attack, however, reached a new, almost absurd, level with Froude presented not just as a charlatan but as the *anti*-historian.

Freeman began by justifying writing about someone who clearly did not qualify, in his mind, as a historian. The historical community, Freeman argued, 'cannot welcome [Froude] as a partner in their labours, as a fellow-worker in the cause of historic truth'. But unfortunately, Freeman admitted, Froude had a popular following and 'will be read by many and will be believed by some'. It was Freeman's duty, in other words, to (yet again) tediously point out the many mistakes that necessarily excluded Froude from the professional historical community.[62]

Throughout his tortuous attack, Freeman suggested many possible reasons for Froude's 'endless displays of ignorance' and 'chronic inaccuracy', besmirching Froude's character and intellect. Initially Freeman chalked up Froude's errors to that of a novice, that 'they are the errors of a man who had taken up historical writing and historical study in the middle instead of the beginning' of his life. 'Constant inaccuracy of reference and quotation betray the man who has begun to write without having gone through any thorough discipline of reading.'[63] Later, Freeman claimed that accuracy for Froude was 'clearly forbidden by the destiny which guides Mr. Froude's literary career'.[64] Not to be outdone by himself, Freeman also connected Froude's failures to a 'fanatical hatred towards the English Church at all times and under all characters'.[65] Freeman was still making Froude pay for *Nemesis of Faith*.

While Freeman suggested a variety of possible explanations for Froude's many problems, in the end he had to agree with the consensus of the historical community who 'are now disposed to set down Mr. Froude's vagaries of narrative and judgement to an inborn and incurable twist, which makes it impossible for him to make an accurate statement about any matter'. In other words, Froude could not help but distort the most basic fact, transforming history into fiction. Freeman was quite clear that this was not his own personal judgement but that of the professional community of historians at large. '[H]istorical scholars', argued Freeman, 'those who have lived and made their homes in the ages in which Mr. Froude shows himself only as an occasional marauder, see ... that when Mr. Froude undertakes one of the simplest of tasks, that of fairly reporting the statements made by a single writer, he cannot do it;'[66] 'The evil is inherent, it is, inborn.'[67] Freeman determined that Froude's 'Life and Times of Thomas Becket' was not only bad history, but that it should not be considered history at all:

> The 'Life and Times of Thomas Becket,' whatever it may be, is not history; because history implies truth, and the 'Life and Times of Thomas Becket' is not truth but fiction. It does not record the life of a Chancellor and Archbishop of the twelfth

century, but the life of an imaginary being in an imaginary age.... History is a record of things which happened; what passes for history in the hands of Mr. Froude is a writing in which the things which really happened find no place, and in which their place is taken by the airy children of Mr. Froude's imagination.[68]

The message, of course, was that Froude, because of an 'inborn and incurable twist', was unable to write history and therefore was unable to be a historian.

As if this were not enough, Freeman claimed that Froude's true intentions behind producing his fictional portrayal of Becket was to damage the memory of his long deceased brother. The second volume of the older Froude's *Remains* was also a history of Thomas Becket and it was precisely the kind of hagiographical saint-worship produced by the Tractarians that so-called Broad Churchman such as Kingsley and Froude disliked.[69] Froude was, according to Freeman, merely trying to challenge his brother's interpretation and stir the pot, as it were, in the same way that he sought to excite readers with his absurd reconstruction of Henry VIII's reign in his *History of England*. What was worse, argued Freeman, was that Froude did the dastardly thing of attacking his brother's interpretation without even naming him thereby pretending as if his much earlier history of Becket did not exist. 'Natural kindliness', argued Freeman, 'if no other feeling, might have kept back the fiercest of partisans from ignoring the work of a long-deceased brother, and from dealing stabs in the dark at a brother's almost forgotten fame'.[70] This dredging up of thirty-year-old history made Freeman's criticisms appear very personal. Froude could not this time ignore them.

Froude responded to Freeman in the pages of the *Nineteenth Century* where his series on Becket first appeared. (He had previously been unable to reply to Freeman's reviews; the *Saturday Review* published Freeman's review anonymously the 1860s, and a short reply sent to the editor at the time 'was refused in language which showed that it would be useless for me to make another application'. Freeman was bold enough to sign his most recent review.) Froude felt that he could not let the occasion go without responding, given that Freeman had clearly 'gone beyond the office of reviewer'.[71]

Froude sought to turn the tables on Freeman in much the same way he had on Smith, suggesting that it was Freeman who suffered from the exact faults he attributed to Froude. 'I think I shall show that "prejudice," "passion," "ignorance," "inability to state facts correctly," "going beyond the evidence," "exaggeration," "an incurable twist," or, as he sometimes puts it, "persistent ill luck," whether they are or are not characteristic of my own writings, have certainly distinguished Mr. Freeman's remarks upon myself.'[72] Froude made this most clear in his discussion of the personal nature of Freeman's attacks, particularly where his brother was concerned. 'How can Mr. Freeman know my motive for not speaking of my brother in connection with Becket, that he should venture upon ground so sensitive? ... Natural kindliness would have been more violated if I had specified my brother as a person with whose opinions on the subject I was compelled to differ.' As to the 'stabs in the dark' comment, Froude replied that 'If I had written anonymous articles attacking my brother's work, "stabs in the dark" would have been a correct expression; and Mr. Freeman has correctly measured the estimate likely to be formed of a person who could have been guilty of doing anything so discreditable.' Froude went on to explain just how much he respected his brother's 'excellences of intellect and character'. He was, 'Irrespective of "natural kindliness," ... the most remarkable man I have ever met'. But Froude admitted to being 'ashamed to have been compelled, by what I can describe only as an inexcusable insult, to say what I have said'.[73] It is an understatement to say that Froude was successful in showing quite clearly the personal nature of many of Freeman's exaggerated and passionate assaults thereby undermining Freeman's self-promoted disinterested and scientific identity. Indeed, Froude's essay does a wonderful job of slaying Freeman with his own sword. Lytton Strachey would later describe Froude's response as 'crushing'.[74]

While Froude's article 'ran rings around Freeman', he may have let his anger get the better of him in his response.[75] He let it be known that he was very upset by Freeman's statement that he had an 'inborn and incurable twist', and he made the mistake of repeating the phrase twice in an attempt to reveal Freeman's vindictive tone.[76] Froude may have 'stood up for himself against libelous criticism' but, as his most recent biographer argues, he 'won the battle for his reputation in a way that left himself open to losing the war'.[77] He did so in particular by circulating discussion about a disease that kept him from being a proper historian. It matters little that he repeated the phrase in order to discredit it; merely by giving the ridiculous criticism the dignity of a response, Froude helped bring 'Froude's disease' into public discourse.

Freeman, however, was incensed by Froude's reply and quickly sent a few sabre-rattling letters to his friends asking advice on how to respond to Froude's clear breach of conduct. The American historian and now Regius Professor of Civil Law at Oxford, James Bryce, suggested that Freeman drop the matter entirely. Aside from suggesting that 'you merely criticize [Froude] as a literary man' and not from 'the ground of some personal hatred', Bryce could not see what possible form a response could take. 'I do not see how you could answer in the controversy' given the public's likely inability, at this stage, to properly judge the matter. If the public had not by now been convinced of Froude's historical forgeries, in other words, they would certainly not be convinced by any future attack.[78]

Stubbs was equally circumspect, suggesting that if Freeman did decide to respond he should only 'answer [Froude] on the Historical points which he has singled out, on which he leaves himself open to criticism'. 'Having done that', Stubbs continued, 'I should say whether I was willing to soften down anything that I had said – If not, then say so, but without iterating anything. ... Then leave it and let the matter drop. You have already amply justified yourself for what you have said, in the judgment of the friends of R. H. Froude.' Indeed, much like Bryce, Stubbs believed that any further response should be short and should avoid bringing up Froude's brother or besmirching Froude's character any further. 'The people who will take their opinion on the matter from their admiration of Froude's conversational powers and their dislike of you or of Truth generally', Stubbs continued, 'will not be convinced by anything that you or anybody else may say'.[79]

Freeman may well have planned to take the gist of his friends' advice and decline to respond, but then he read the 5 April 1879 issue of the weekly *Spectator* and saw that Froude's 'A Few Words on Mr. Freeman' had gained sympathetic notice, much of it at Freeman's expense. The *Spectator* clearly sympathized with Froude's plight and argued that despite Freeman's invidious assaults, Froude 'displays a really admirable command of temper, and great fidelity of gentle retort'. The *Spectator* believed that Froude was able to show 'conclusively that he does take trouble over his work, and give time to it, which Mr. Freeman had denied.' Froude's response would 'definitely raise the public impression of his personal character'.[80] This Freeman could not have.

In his 'Last Words on Mr. Froude', Freeman used the *Spectator* article as an excuse to answer Froude's main objections as well as to reproduce the well-worn criticisms directed against Froude over the years. Freeman denied that his twenty-year assault on Froude was somehow personal, claiming that 'I know nothing about his personal character. Mr. Froude is to me simply the writer of certain books. Whatever I have said about him has arisen naturally from his writings.' This, of course, was the rub. 'I believe those writings to be, in more ways than one, misleading and dangerous, and I have spoken accordingly.'[81] In other words, Freeman's tone may have appeared offensive to a reader of the *Spectator*'s defence of Froude, or of Froude's *Nineteenth Century* article, but any reader of Freeman's reviews of Froude would know what was at stake with Froude's histories and would therefore understand Freeman's seemingly violent tone.

Freeman attempted to heed Stubbs's advice to stick to the historical objections raised by Froude, but this took the controversy to the absurd level Bryce wanted Freeman to avoid. Freeman quoted from Froude's response, from Froude's 'Life and Times of Thomas Becket', from his reviews of that work, from Froude's *History of England*, and from his many reviews of that work. His purpose in dredging through twenty-year-old works and reviews was to show the reader of the *Spectator* or of Froude's 'A Few Words on Mr. Freeman' that he was justified in his assaults and was more than fair. These points of controversy were also derived from Froude's response, and Freeman was clearly delighted to show that even in this situation 'Mr. Froude is pursued by his usual ill-luck – by that hard destiny which makes it impossible for him accurately to report anything.'[82] Just as Froude's historical works are filled with inaccuracies, argued Freeman, so are his attempts to defend his work in the face of attack. 'As he misquotes his authorities, so he misquotes me; nay, he goes a step further still; "in seipsum postremo saeviturus, sit cetera desint," he misquotes himself.'[83]

Freeman was extremely proud of his 'Last Words on Mr. Froude' believing that he had finally – and without question – slain his enemy. 'Have you read my answer to Froude?' he wrote to Bryce; '[s]urely you allow now that it was right to answer such a litany of misrepresentations – What a queer creature [Froude] must be.'[84] He also wrote to his friend, J. R. Green, convinced that his last response would finally persuade 'all reasonable people' of Froude's 'strange mental and moral twist'.[85]

Eventually Freeman's boundary work paid off as he succeeded Stubbs in 1884 as Regius Professor of Modern History at Oxford when Stubbs moved on to take up a position in the Church hierarchy as Bishop of Chester and then of Oxford (1888). Unfortunately much of the shine had been taken off of the Regius trophy by then. Freeman had heard many of Stubbs's complaints over the years. Almost immediately upon getting the post, Stubbs admitted that he did 'not believe that anybody at Oxford cares about History except as a matter of education. I suppose it is alright but it takes the shine off things.'[86] Of course, it was Stubbs's main task to change just that environment but he was convinced the best students were kept from him and he complained bitterly about his administrative duties and the writing of lectures.[87] Stubbs even complained in the preface to his published lectures about

> [t]he feeling of compulsion, the compulsion to produce something twice a year which might attract an idle audience, without seeming to trifle with a deeply loved and honoured study, was so irksome that never once, in the course of my seventeen years of office did I think that there would come a time when I could look back on this part of my work with pleasure or grateful regret. And I fear that this will be only too obvious to anyone who tries to read this book.[88]

He admitted that his last lecture was a 'matter which mingles pleasure with pain, in no slight measure of both' and that perhaps 'he never was fit for the place [Oxford]'.[89] It should not be surprising that Stubbs's lectures were poorly attended.[90]

Freeman was certainly happy to finally achieve the post he had coveted for so long, but it was bitter sweet. In a letter to Goldwin Smith, Freeman said that he was honoured to succeed Stubbs, Smith and Thomas Arnold before him, 'but I gnash my teeth that I have not had you and Stubbs to my colleagues, and not to predecessors'. He confessed that '[y]ears ago to fill one of the historical chairs at Oxford was my alternative ambition with a seat in Parliament. It seemed for years as if neither would ever come to me: and now at last one has come when I am rather too old for change.'[91] He, too, complained about his duties as chair and just like Stubbs, his lectures 'were delivered for the most part to very empty benches'.[92]

Freeman would have been happy to know, however, that his boundary work against Froude had an unintended consequence. His endless criticisms of Froude were picked up by the Frenchmen Charles Victor Langlois and Charles Seignobos who wrote a manual on the proper scientific study of history. They held up Froude as the ultimate example of a writer of history unable to live up to the scientific ideal because of his countless inaccuracies. '[I]t has often been said of Froude', Langlois and Seignobos claimed, that he must have suffered from a disease that kept him from saying anything that was accurate. They went on to diagnose 'Froude's disease' as simply 'incompatible with the professional practice of historical scholarship'.[93] In 1898 the book was published in English as an *Introduction to the Study of History* complete with a preface written by then Regius Professor of Modern History in Oxford, F. York Powell who described the book as a must-read for students of history as well as for professional historians because it presented 'history as a scientific pursuit' and promoted the importance of a scientific methodology.[94] The French manual would become well-known in the English-speaking world, where Froude's disease would become a common phrase used to invoke an extremely faulty historical method, particularly in scientific research manuals throughout much of the twentieth century, in order to point out that those suffering from the disease 'are not qualified to engage in a critical investigation'.[95] Perhaps such would have consoled Freeman had he known that the original sufferer of Froude's disease would succeed to his Oxford chair of history in 1892.

5 HISTORY FROM NOWHERE

Do not image you are listening to me, it is history itself that speaks.

Lord Acton (paraphrasing Fustel de Coulanges), 'The Study of History' (1895)

I am very convinced that the one who drives the coach should commit himself to nothing.

Mandell Creighton to James Bryce, 28 December 1885

Establishing history as a science certainly seemed like violent sport, largely promoted through attacks and arguments about what history should not be, rather than what it could and should be. There was a fairly consistent message behind the attacks, however, one that suggested the historian must suppress his subjectivity and present the facts in an inductive fashion. For Seeley, this involved understanding that history is not dramatic, that it is not inherently structured like a novel. One of his main goals over his tenure as Regius Professor was to impress on students and the general public that history had to eschew its literary past and embrace a way of thinking and writing about the past that likely would not be fascinating to the ordinary reader. Freeman was less convinced that history could not be entertaining but he also knew that there was a temptation to make it more entertaining than it already is with a twist of phrase or simply a literary style in general that might force a weak mind such as Froude's to consistently skew the facts to fit the story rather than making the story fit the facts. There was something very moral about this kind of criticism that suggested the historian had a duty to avoid the obvious temptations that writing a popular book would entail and merely present the past as it actually happened. Scientific history took a fair amount of self-discipline and weak minds such as Kingsley and Froude were simply not up to the task.[1]

What was missing in the boundary work done by the likes of Seeley, Freeman and Stubbs was a more positive message about the inductive science of history, one that made it seem more attractive while highlighting the traits promoted by the historical community of self-discipline, of a strong work ethic, of a painstaking accumulation of particular facts, of a narrative devoid of style. Lord Acton, who returned to active historical duty in the 1880s and 1890s, was primarily responsible for helping to communicate the more positive attributes of this method and much like Ranke before him he was able to convert the various principles that scientific historians appeared to rely upon into a seemingly simple paradigm that centred on a broad conception of objectivity.

When Acton was picked to replace Seeley upon the latter's death as Regius Professor of Modern History at Cambridge in 1895, the appointment was largely interpreted as a watershed for both Cambridge and the study of history in Britain. Acton's reputation clearly preceded him and while there were some detractors questioning the appointment of a man who had not completed a detailed study of a specialized historical subject, great things were largely expected from most observers.

After lamenting Cambridge's choice of Regius professors in comparison to that of Oxford, the *Speaker* argued that this time the ancient university had finally got it right. The *Speaker* highlighted Acton's legendary learning and lack of general attachment to schools of thought or university cliques, arriving at 'Cambridge with a ripe knowledge and a trained experience, with no bias except a moral bias, without insularity or religious prejudice, to stimulate, we are sure, and also, as we hope, to organize and direct'. The *Speaker* had no doubt that Acton would share with his students his international learning, particularly 'the German's power of collecting detail and the Frenchman's power of assimilating it, with the German's knowledge of facts and the Frenchmen's love of generalization and of phrase'. Indeed, the *Speaker* was greatly impressed that Acton 'combines a sturdy English common sense which prevents him being the slave of any theory'.[2] Even the *Saturday Review*, a publication that had not been kind to Acton's predecessors, found the appointment 'not an unpleasant surprise', in part because of Acton's previous exclusion from the university, a point also made by the *Speaker*. As well, the *Saturday* held out much hope for history at Cambridge under Acton's leadership given 'the fact that he had sat at the feet of such masters as Ranke and Döllinger'. There was much hope that 'something out of the common might be expected' from Acton's appointment.[3]

W. S. Lilly, writing for James Knowles's *Nineteenth Century*, believed Acton's appointment to be 'among the most important events that have for a long time occurred in English academic life'. The fact that an ancient university had so clearly welcomed a previously excluded Catholic with open arms was wonderful evidence of 'the passing away of that old sectarian spirit which found expression in religious tests'. But there was another reason, Lily explained, why Acton's appointment was so special. 'He is, beyond all question, our most learned representative of the modern spirit in history – the scientific spirit.'[4] Acton, according to Lily, was simply in another category of learning than all other previous holders of the office. Unlike his predecessors, '[h]is laborious life has been devoted to historical research, pursued in true scientific methods'.[5] Expectations for his inaugural lecture were certainly high and for the most part he did not disappoint.[6]

Acton began his inaugural lecture by looking back 'to a time before the middle of the century', where he himself 'applied for admission' to three colleges at Cambridge but was unfortunately 'refused by all'. Being a Catholic would have all but barred Acton from attaining an undergraduate degree when he was a young man. But things had clearly changed at the ancient universities. In the forty-five years since Acton first applied, the universities had been transformed. The training of future clergymen was no longer the universities' primary function; it followed that both student and teacher need not be expected to adhere strictly to Anglican articles of faith. Indeed, graduates were no longer required to subscribe to the Thirty-Nine Articles. It was in this way that Acton's appointment was a symbol of the declining role of the Anglican Church at Cambridge.[7] 'Here from the first', Acton addressed those in attendance, 'I vainly fixed by hopes, and here, in a happier hour after five-and-forty years, they are at last fulfilled'.[8]

Acton did not dwell on the former exclusionary practices of Cambridge colleges, however, but instead gave a sweeping history of the teaching of modern history at Cambridge. He discussed the slow but steady development of history as a professional discipline throughout the century, appearing to reach some sort of climax coinciding with his own appointment at Cambridge. Acton sought to explain how the study of history had so impressively improved over the century, how the practice became more authoritative, more scientific, more detached. He paid homage to the by now acknowledged founder of the modern profession of history, of history as a truly scientific study, 'my own master' Leopold von Ranke.[9] 'Ranke is the representative of the age which instituted the modern study of history.' Surely there are more interesting studies produced by much more creative imaginations but as a detached observer, as a methodical practitioner of history as a science, 'he stands without rival'. More than anyone, he taught that history ought 'to be critical, to be colourless, and to be new. We meet him at every step, and he has done more for us than any other man'.[10]

Here Acton was echoing a familiar dictum of Stubbs, that the scientific historian need not be a genius, but a diligent and mechanical, even 'colourless' worker. And this was what, according to Acton, separates the new scientific history from romantic, popular, or even positivist history. 'The strongest and most impressive personalities ... like Macaulay, Thiers, and the two greatest of living writers, Mommsen and Treitschke, project their own broad shadows upon their pages.' Such men were engaged in their own particular projects which required an immeasurable amount of knowledge to complete, but such studies were overly burdened by the very presence of the author. 'This is a practice proper to great men', argued Acton, but for history to be a scientific and systematic analysis of the past it could not be analysed by great men who overburden their subject matter. For Acton, the historian was at his most authoritative, at his historical best, when he not only cast no shadow, but when he disappeared into his own narrative as if it were history itself that spoke.

> [T]here is virtue in the saying that a historian is seen at his best when he does not appear. Better for us is the example of the Bishop of Oxford [Stubbs], who never lets us know what he thinks of anything but the matter before him; and of his illustrious French rival, Fustel de Coulanges, who said to an excited audience: 'Do not image you are listening to me, it is history itself that speaks'.[11]

We must, Acton explained, move beyond the 'historians of former ages' who are so 'unapproachable for us in knowledge and in talent'. They 'cannot be our limit. We have the power to be more rigidly impersonal, disinterested and just than they'.[12]

For the historian to be truly disinterested, Acton believed that he must overcome his subjective self and transcend that temptation to falsify in order to be true to his sources. There is certainly significant overlap between this concept of objectivity and the mechanical form of objectivity which Lorraine Daston and Peter Galison claim was dominant in the sciences in the second half of the nineteenth century. They argue that there was a widespread belief at the time that held that in order for scientific knowledge to be both created and disseminated to the wider scientific community the scientist sought to suppress all individual perspectives in order to let nature speak for itself.[13] Acton and other Rankean historians certainly embraced this form of objectivity, but for Daston and Galison there was something passive and timid about mechanical objectivity that sought expressly to eschew judgement of any kind. For Acton and other Rankean historians such as Seeley and Freeman, however, suppressing one's subjectivity was not the same as suppressing judgement. They sought, rather, to overcome petty self-interest and Romanticizing temptations precisely so that they could be trusted to impart not personal subjective interpretations but the truth. This required not just the proper method but also a great deal of self-discipline, a process for which the term 'mechanical' does not quite do justice. In this way the Rankean historians were very much like men of science such as Thomas Henry Huxley and Karl Pearson in the sense that they sought to construct disinterested selves who could be trusted to impart knowledge that transcended subjective perspectives. As Theodore M. Porter explains, this form of objectivity 'was mainly a positive moral trait', one that sought not to 'annihilate judgment' but to exalt it.[14] Acton made this clear enough when describing the often 'neglected truth' of historical justice. He exhorted his audience 'never to debase the moral currency or to lower the standard of rectitude, but to try others by the final maxim that governs your own lives, and to suffer no man and no cause to escape the undying penalty which history has the power to inflict on wrong'.[15]

For Acton, the scientific historian had a moral duty to be 'rigidly impersonal, disinterested *and* just', an identity that was not in Acton's mind contradictory.[16] But it was certainly not an identity that was possible for weak historical minds. He wanted historians to express their views, but from within a disinterested discourse, as if they were expressing, in the great phrase of Thomas Nagel, a 'view from nowhere'.[17] The man of scientific history, much like the man of science, had ideally to enter an interstitial space between presence and absence, between existence and non-existence, in order to investigate and disseminate the truth of the past.[18] As Lord Acton explained, the truly detached scientific historian would exhibit 'what Michelet calls *le désintéressement des morts*'.[19]

Acton was not saying anything terribly new about the science of history as the extensive notes to his published lecture make clear but he did seem to be saying what was already quite well known and practiced in a new way. Part of the problem with the inductive approach to history was that it seemed to advocate making history a fairly boring summary of things that happened. In commenting on Acton's lecture, the *Speaker* seemed to understand this all too well, making the point that the reason the Romantic brand of history has endured is because '[i]t is so very easy for the worshipper of heroes, even clay heroes, to collect an audience', while 'the just man who weighs and finds nothing perfect' has a profoundly difficult time just 'getting people to listen'. Indeed, the '"rigidly impersonal" rarely supplies the enthusiasm necessary for creative work; the critic usually knocks down more than he builds up'. However, the *Speaker* believed that Lord Acton's approach might just 'prove the exception to an unhappy rule' as he seemed to show 'us how to be "rigidly impersonal," without failing to be interesting'.[20] Perhaps scientific history could be exciting after all.

It should be made clear that Lord Acton did not just stumble onto the scene in 1895 and suddenly make these well-received pronouncements about the practice of history. British historians were well aware of Acton's immense learning and background as a student of the great German school of scientific history well before his inaugural lecture made that apparent to Cambridge and the public at large. He was also before this time well known as an impartial judge and as a great defender of justice and academic freedom against, in particular, religious persecution. His struggles as a liberal Catholic against the Church's more conservative hierarchy were well documented in the *Rambler* and its quarterly successor the *Home and Foreign Review*, both edited by Acton. He most notably refused to be dictated to by the Catholic Church and eventually shut the *Home and Foreign Review* down rather than succumb to the pressure of Rome, a move that was noticed by non-Catholic observers.

For many years before Acton would officially take up a post in guiding the historical profession he had acted as somewhat of a spiritual advisor for scientific history-writing in Britain, most notably in the establishment of the *English Historical Review*.[21] It had long been understood that for history to be a professional and scientific discipline of study a journal published by and for professional peers was an important step in this regard. What was perhaps more important was that the establishment of such a journal was deemed a strategic way for historians to communicate their work with one another without having to try and please publishers beholden to a much larger reading public. A proper historical review represented the ideal hopes of a professionalizing community looking for a space for their scientific work. It should not be surprising then that when the journal was first envisioned by J. R. Green, James Bryce and Edward A. Freeman, the exclusion of Romantic history-writers was one of the main concerns, almost twenty years before the journal would finally appear. They wanted the journal to present historical research and not become the venue for Romantic writers to spout their subjective views without the appropriate learning.[22] Despite its envisioned exclusionary nature, Alexander Macmillan only seemed interested in publishing such a journal with Green as editor, in part because Green seemed to understand that the journal could not succeed should it be for historians alone, that, in Green's words, the review should avoid becoming one of 'the special class of Humdrums'. '[P]roperly edited', Green wrote to Macmillan, the historical review 'might be at once a representative of historical investigation, and at the same time have a current and general value for the world without'.[23] When Green's health began to fail, however, so too did Macmillan's interest in the journal. It was not until the early 1880s under the leadership of Mandell Creighton (1843–1901) that the review finally saw the light of day.

Creighton attended Oxford just at the moment when history was undergoing a transformation and new ideas of scientific history were coming to the fore. He began his residency at Merton College in 1862 and eventually took a first class in *Literae Humaniores* (often referred to as the Greats programme) in 1866, the same year Stubbs began his Regius professorship there. Creighton was quickly elected to a Merton fellowship and was tutoring undergraduate students by 1867. Throughout the 1870s he began researching the history of the papacy just before and during the Reformation, an immense contribution to the subject that would eventually be published as *A History of the Papacy during the Period of the Reformation* in five volumes (1882–94). The first two volumes would appear in 1882 and by 1884 he was rewarded for his diligent historical research by being appointed as the first Dixie Professor of Ecclesiastical History at Cambridge. He would later, just like Stubbs, abandon his university chair in order move up the Anglican Church hierarchy becoming Bishop of Peterborough in 1891 and then Bishop of London in 1897 until his death in 1901.[24]

Creighton and Acton would become correspondents and eventually close friends and colleagues following Acton's review of the first two volumes of Creighton's *History of the Papacy* in 1882.[25] Creighton had actually been an admirer of Acton's knowledge of ecclesiastical history and even requested that the editor of the *Academy* ask that Acton review the volumes. 'I specially asked the Editor to get you to review it', he wrote to Lord Acton, 'as I wanted to be told my shortcomings by the one Englishman whom I considered capable of doing so'.[26] He certainly could not have been disappointed on that front. Despite Acton's many criticisms, he did admit that the study was 'the best History of the Reformation' yet produced and he believed that Creighton told the 'history of increasing depravity and declining faith, of reforms earnestly demanded, feebly attempted, and deferred too long ... with a fullness and accuracy unusual in works which are the occupation of a lifetime'.[27] This was high praise from such a severe historical reviewer and Acton's only real criticism, though it was explained and expanded upon throughout the review, was that Creighton was, in places, far too detached from his controversial subject matter. 'His suggestive brevity and dislike of emphasis, his sobriety and reserve, his carefulness to stumble over no problems, to enforce no moral and improve no text, sometimes raise a wish for deeper furrows and closer grasp.'[28] While this general criticism would represent a deeper schism between the two historians' conceptions of objectivity and foreshadow a more divisive debate to come, Creighton admitted that he was perhaps too conscious of being a 'Protestant writer, with the best of intentions, to be accurate in his reading and interpretation'. In this sense, the 'need for comprehension is the difficulty which I find at every step'.[29] Acton responded to Creighton's letter very graciously while not backing away from any of his criticisms and Creighton acted the part of the student who had just been chastised by the acknowledged master, thanking him and asking for more.[30]

When Creighton with the help of Bryce convinced publisher Charles Longman to take a chance in publishing a quarterly journal of specialized historical research and reviews, it was to Lord Acton that Creighton turned for guidance, contributions, as well as help in establishing an editorial statement of policy. Creighton, unlike Green, was at least early on fairly consistent in his message that the journal was primarily for historians and must present papers written from the scientific perspective. It was for this reason that Creighton excluded Froude's name from the publisher's list of possible contributors.[31] He also ensured that the major proponents of scientific history would publish in the journal's first issue, and he was able to secure contributions from Freeman, Seeley and Samuel Gardiner. Creighton was adamant, however, that Acton contribute something. Given that most of his authors were names well known and have 'already had their say' in other publications many times over, an article from Acton 'would attract attention at once'.[32]

Acton suggested writing on several topics and only at the end of a very long letter posited the possibility of writing a review of a recent book on German historiography by Franz von Wegele. 'If you have not a specialist for it, some of the Germans would be on the look out for a notice of it by me, as having lived among people of this sort'.³³ After not hearing back from Creighton and realizing that the significance of reviewing Wegele may have been unclear, Acton wrote again petitioning Creighton 'to review Wegele's history of German literature; and suffer me to renew my application and to underline it'. Since his previous letter, Acton came to realise that if he could be allowed to 'notice this book in my own way, that is to say, not at all, but to give my own account of the schools and events, the merits and failings of the Germans' he might be able to provide the more substantial article Creighton seemed to prefer, albeit framed as a review. Acton was essentially suggesting to write an analysis of German historiography from his own perspective, while using Wegele's book as a hook to write such an analysis. This way as well, argued Acton, 'It would be less didactic, less presumptuous by reason of the smaller type, and much less likely to offend Oxford and Cambridge'. Acton even hoped that perhaps Creighton could 'hide it away at the end of your literary notices, a place variously appropriate, and allow me, considering the occasion, indefinite pages'.³⁴

Creighton was delighted by this suggestion and it sent his mind reeling at the possibilities of Acton writing something else for this issue as well. 'I should rejoice in a review of Wegele' argued Creighton, but he went on to suggest that Acton should make extensive 'introductory remarks' for the journal as a whole, 'not only about Reviews, but about the function of history generally'. Creighton was essentially asking Acton to write the journal's inaugural introductory statement, as if on behalf of the English profession of history at large, tenuously entitled by Creighton as 'On the Present Position of Historical Studies'. Creighton admitted that with the exception of Stubbs, who declined to write for the first issue citing ecclesiastical business, 'you are the only person ... who has any claim to guide and direct and the mere fact that you are neither of Oxford nor Cambridge prevents the possibility of offense'. He wanted to avoid asking either Freeman or Seeley because 'everybody knows what [they] think of historical study: the utterances of one would give offense to the other – you alone could speak'. He went on to explain to Acton just how important he was to the journal's future. 'You see more and more am I affirmed in my original impression, that you are the prop and stay of the Review.'³⁵

Acton seemed to be taken aback by this suggestion and it is clear that Creighton misunderstood Acton's rationale for wanting to bury his review at the back of the issue. 'It will not do for me to write an inaugural', Acton wrote to a disappointed Creighton. It was, quite simply, a 'question of propriety and proportion. You would incur a semblance of dilettantism if a man who never wrote a book rushed in where the Angel of Chester [Stubbs] fears to tread.' This was precisely why Acton wanted his views tucked away in a larger review article in the back of the journal. He assured Creighton that he could get his 'knife into every joint without being felt, if you allow me to review Wegele'. Acton believed that Wegele's study left many gaps for him to fill, particularly the 'new lights' of German history-writing. 'There are some 40 or 50 men who have innovated, who have enlarged the definition of history, whom it would be worthwhile to commemorate.' In doing so, Acton assured Creighton that he would 'lay out a chart' for the review and the English historical profession, 'and to convey, innocently enough, a multitude of hints to the unwary reader'.[36]

Despite Creighton's disappointment with Acton's decision, he was delighted by Acton's first draft, even though it was very late and far too long, his original thirty pages had to be cut down to twenty.[37] 'Your article fills me, if possible, with greater admiration than before – How can you know so much? ... Never before has such a prospectus been made, and I do not suppose that anyone except yourself could have made it.' Creighton was sure the article would 'secure the Historical Review respectful attention throughout Europe' and he set about making the article look less like a review as possible.[38] He proposed giving the review a title, 'The Historical Literature of Germany', though Acton preferred 'German Schools of History' in order to highlight the fact that the piece was about 'the growth of a specifically German view of history', while relegating the listing of Wegele's book 'to the bottom of the page', and by that he meant making the review heading a mere footnote.[39] He also somehow convinced Acton that the review should run alongside the articles rather than be hidden away in the review section. Not only that, the article would appear, tellingly, as the first article in the first issue of the *English Historical Review*. If Acton would not write the introductory remarks, Creighton would simply make it appear as if he had.

Even though Acton refused to write such an introduction, Creighton had Acton's review pencilled in as 'On the Present Condition of Historical Study' and seems to have had little interest in writing an introductory statement himself.[40] Bryce, who was the assistant editor, clearly felt an editorial statement was necessary, though Creighton seemed to think that Acton's 'prospectus' might be enough. He was very clear that he was not about to write anything for the first issue. 'I have an editorial caution about writing myself', he wrote to Bryce. He was concerned about offending possible contributors: 'My only object is to do the best I can and advance tentatively.' But, he explained, should Bryce want to 'write some introductory I shall be only too glad' for it.[41] Perhaps Creighton felt that he, too, should not proceed where Acton and the 'Angel of Chester' feared to tread, though it seems odd that he would be so willing to let the assistant editor complete the task.

James Bryce (1838–1922) was no editor's lackey, however. Not only was he Creighton's senior by five years, he was from an outspoken family of highly educated Ulster Scots. His father was a geologist and his love of natural history transferred to his son who attended Glasgow University in 1854 before moving on to Trinity College, Oxford in 1857 thanks to a scholarship. Like Creighton he went through the Greats curriculum graduating first in his BA class in 1862. As a nonconformist, he was unable to complete an MA. This did not stop him from becoming a great scholar of European politics, law and history, writing for several periodicals throughout the late 1860s and 1870s.[42] Bryce had written a prize-winning essay on the Holy Roman Empire, an essay that greatly impressed Freeman, and the two became very close friends. Consequently, Bryce was drawn into Freeman's circle of friends that included in particular Stubbs and Green. His essay was eventually expanded into book form and became the standard Victorian study of the empire. He would also write substantial studies of legal history and of the United States. In addition, he passed the bar exam and practiced law in London for several years.

Bryce was an integral member among this group of scientific historians. He was involved in just about every attempt to establish an historical review and he tended to agree with Freeman when it came to punishing historical scholars who did not live up to the appropriate standards. Interestingly, it was his work as a law scholar, rather than primarily his historical scholarship, that secured his appointment as Regius Professor of Civil Law at Oxford in 1870, a post he held until 1893 when he became more involved in politics. He would later become England's Ambassador to Washington from 1907 until 1913. When he agreed to help Creighton get the *EHR* started it was generally as an equal, though he did not have the title co-editor and Creighton's correspondence with him is startling when compared with that of his correspondence with Acton. For Creighton, Bryce was simply not the calibre of scholar that Acton was. But he was more than willing to let Bryce write the inaugural editorial statement, if only to avoid doing so himself.

Creighton was very grateful for the editorial statement when it was in his hands only a week later. He thanked the 'angelic' Bryce for writing it and was thoroughly pleased by it: 'What you have said seems to me excellent, and I could not have said it so straight forward.'[43] He tried to again explain why he was so 'glad not to write a preface' himself and claimed it was because he 'knew too much about the views of too many people', those who will think the journal too dull, or not scientific enough, those who will think the journal is too modern, or not continental enough, etc. But this way, should anyone complain, he can claim that he did not write it. 'I am very convinced that the one who drives the coach should commit himself to nothing.' He also suggested that the editorial statement appear 'anonymously' and the only change Creighton himself made was to alter the final sentence.[44]

Bryce may have written the 'Prefatory Note' but by leaving it unsigned, it appeared as if penned by the historical community at large, acting as a communal editorial statement. Perhaps this was the effect Creighton was hoping for. It was only six pages long but they tell us a lot about how Bryce, and to lesser extent Creighton, expected the journal to function within the community itself. After lamenting the fact that England was 'alone among the great countries of Europe' where 'there does not exist any periodical organ dedicated to the study of history', Bryce went on to say that thankfully such a thing is no longer true. It only makes sense that England would have a review devoted to historical research and reviews given that the work done there 'is as thorough in quality as that even of the Germans' and, of course, it was larger in quantity and likely better in quality as well than the French and Italians.[45] Bryce continued that the need for the journal was clearly evident and it was finally possible 'to focus the light now scattered through many minor publications, none of them devoted to this special purpose, which shall present a full and critical record of what is being accomplished in the field of history, and become the organ through which those who desire to make known the progress of their researches will address their fellow labourers.'[46]

Bryce explained that the principles by which the journal would be guided, as well as the methods that would be promoted, would be best illustrated within the contents of the journal itself. In other words, the articles themselves would be the best explanation of the editorial principles and standards and they would be the best explicators of the proper historical methods rather than any editorial statement of such principles. He admitted, however, that there were likely some preliminary questions that would need to be answered at the outset and he proceeded to explain some of the key principles of professional history-writing and their centrality in shaping the editorial policies of the journal.[47]

One of the questions Bryce asked himself had to do with partisanship and how the journal would avoid appearing to support, for instance, particular political points of view or ecclesiastical persuasions. 'It will avoid this danger', argued Bryce, by simply 'refusing contributions which argue such questions with reference to present controversy'. This would likely not be much of a problem though, Bryce explained, for the 'object of history is to discover and set forth facts' and the historian therefore 'confines himself to this object' and 'can usually escape the risk of giving offense'. He admitted that perhaps some topics would be unavoidable at which point both sides would need to be given equal weight. 'But our main reliance will be on the scientific spirit which we shall expect from contributors likely to address us.'[48] Contributions not addressed to such a scientific spirit would simply not be considered.

The final question Bryce posed himself was about the audience the review would address: to 'professional students of history, or to the person called the "general reader"'? To this question, Bryce was unambiguous. The review would address the former. 'It will, we hope and intend, contain no article which does not, in the Editor's judgment, add something to knowledge, i.e. which has not a value for the trained historian.' The journal would simply not accept unoriginal or popularly addressed work. 'No allurements of style will secure insertion for a popular *réchauffé* of facts already known or ideas already suggested.'[49] This was a fairly blatant shot taken at a form of history that professed to be artistic rather than scientific, perhaps primarily directed at Froude, a historian who would not be welcome to publish in the journal's pages.[50]

While Bryce's editorial statement made it fairly clear that the *English Historical Review* would be quite exclusively for scientific history, Acton's review of Wegele provided an impressive critical survey of German history-writing in the nineteenth century, and therefore an illustration of the trajectory of England's borrowed scientific method of history. Acton's survey, however, was by no means the work of historiographical hero-worship. He made it quite clear to Creighton that he would highlight both the positive and negative attributes of recent German historiography. 'Everybody will suspect me of overpuffing my own masters', he wrote to Creighton, 'so that I want to make use of my opportunities whenever the superiority is not on the side of the Germans'. Here he was suggesting that English history-writing was getting better in certain respects and he certainly did not want it to 'appear that we lack men who can tackle the great questions of central history'.[51] As Acton explained in the article itself, he set out to show neither the 'infirmity' of German historians, 'nor their strength, but the ways in which they break new ground and add to the notion of the work of history'.[52]

'German Schools of History' was an impressive piece of scholarship by someone who clearly knew the twists and turns of his subject matter. Acton focused his attention on the central roles of Niebuhr and Ranke, of Mommsen and Sybel, of Droysen and Treitschke. He described the initial influence of Romanticism and Hegel, of nationalism and eventually imperialism, and of science and the uphill climb begun and continued by Ranke throughout the century against men of genius who were unable to grasp the much more mechanical processes necessary to make use of archival sources and the discipline required to collect, condense and interpret the masses of new materials. More impressive still was Ranke's ability to maintain impartiality, not just when analysing documents, but when analysing character, a much more complicated task indeed.[53] Much of the views later expressed in his inaugural lecture were expressed here, though not quite as eloquently.

Creighton thought the article 'admirable' and told Acton that it 'has attracted much attention in England'. Creighton even suggested that Acton 'enlarge it considerably' because some were simply unable to follow many of his allusions and were unfamiliar with a few of the lesser historians critiqued.[54] This was Creighton's subtle way of suggesting that there were many parts of the essay that were quite ambiguous. This would become a familiar critique of Acton's work. Creighton was much more forthcoming to Bryce about his opinion: 'Acton's article is admirable; but like him in all things leaves in doubt what is his own point of view.' Creighton was certainly impressed at the way in which Acton's 'attitude of general detachment is skilfully managed so as not to slip into general superiority', a trait he clearly wished he could possess given his inability to pen an inoffensive editorial introduction, but he also had to admit that such a profoundly detached style can at times be 'bewildering'.[55] Creighton was here getting at the inherent double-bind that one encounters particularly in Acton's attempt to exclude himself from his writing. Letting the past itself speak was by no means an easy task and Acton's meaning was often lost from his analysis, one that so clearly bore the markings of immense learning, but whose finer points about German historiography were lost on many. For example, claiming that Acton's article 'exhibits very well his enormous erudition and his subtle and interesting thought', the *Saturday Review* argued that the piece was unfortunately 'marred here and there by eccentricities of expression'. The *Saturday*'s opinion of the review as a whole, however, was that it was a 'capital number'.[56]

Acton believed that the inaugural issue was largely a success. He congratulated Creighton 'sincerely', stating that the 'Review is solid, various, comprehensive, very instructive and sufficiently entertaining'. If he had to compare it to the other historical journals he would place it below Germany's *Historische Zeitschrift* but above France's *Revue Historique*. Acton was impressed that 'at least half the names are there' and more importantly was able to 'discern the makings of a Sacred Band of University workers. There is no reason why it should not become, by the end of the year, the best of all historical reviews.' He was, however, not uncritical pointing out some fairly prescient problems the journal would continue to exhibit for many years. It was not, Acton explained, '*Bahnbrechend*' (groundbreaking). It also 'makes no striking discovery and does not exhibit a great new force, or open a new vista'. And most worrisome indeed: 'it produces no new men'.[57]

Acton would continue acting as Creighton's close advisor on the *EHR* for many years to come, quite regularly suggesting articles, reviews, possible contributors and general advice on how to improve the journal. When volumes three and four of Creighton's *History of the Papacy* were published in 1887, he could not think of anyone better than Acton to review them. 'Will you review them for the Review?' Creighton wrote to Acton. 'I ask you because it would be a great kindness to me if you would do so, and secondly because I know no one else who would believe me if I told him that I was thankful for criticism and really had a very poor opinion of my own productions.' Creighton was insistent that his 'editorial functions' would not get in the way of Acton writing a critical review and he hoped that Acton 'will knock me about the head as I deserve' as Acton had also done with Creighton's first two volumes just five years before.[58]

Acton gladly accepted and he certainly did not hold back in knocking Creighton 'about the head'. His very general criticism of Creighton's earlier volumes was reproduced ten-fold in his current review. He wrote to Creighton frankly suggesting that as editor, he should probably get a few others to read over the review as it was extremely critical. 'You must understand', he wrote to Creighton, 'it is the work of the enemy'. Acton came to realize what Creighton must already 'partly know' – that there was a 'yawning difference between your view of history and mine'.[59] Of course Creighton was unaware of this 'yawning difference' between their views but the fascinating exchange that followed indicates that perhaps there was a general misconception between inductive scientific historians about what it actually meant to let the past speak for itself and the role of moral judgement within such a generally detached view.

Creighton's volumes dealt with the period of the Protestant Reformation and the Catholic Counter-Reformation and he was very careful to avoid passing too strong a judgement on the actions of the papacy should he be criticized for merely providing the biased account of a Protestant historian. Creighton sought to detach his personal opinions from the narrative and let the facts speak for themselves – thereby following the Rankean method to the letter. Acton argued that Creighton was actually doing a disservice to 'historical accuracy' by seemingly ignoring the most horrific elements of the Inquisition even denying 'at least implicitly, the existence of the torture-chamber and the stake'. From Acton's perspective, Creighton was actually shirking the responsibility to determine right and wrong, even ignoring 'the ordinary evidence of history'.

For Creighton, at issue was whether or not the historian, with the benefit of hindsight and influenced by his own personal conceptions of right and wrong, had a right to pass judgement on events of the past. Surely this was up to the reader to decide assuming that all of the facts are presented in a thoroughly detached manner. But for Acton there are concepts of right and wrong, good and evil, that transcend the vast terrain of history and even personal biases. There is a universal Christian morality, Acton had to remind Creighton, that acts as the ultimate arbiter and it is the historian's duty to uphold that moral code when narrating the events of the past. 'Historic responsibility has to make up for want of legal responsibility', Acton explained.

Relying on this moral code in an historical analysis does not undermine history's scientific status, Acton argued; in fact, it is entirely necessary for the historian to be truly objective. The most important aspect of the Christian 'moral code' is its 'inflexible integrity' and that is 'the secret of the authority, the dignity, the utility of history'. Without relying on such an inflexible moral code, 'History ceases to be a Science, and an arbiter of controversy, a guide to the wanderer, the upholder of that moral standard which the powers of earth, and religion itself, tend constantly to depress.' This was a 'plainer and safer' method, argued Acton, than one that refuses to make historical judgements on the basis of historical relativity. 'Power tends to corrupt', argued Acton, 'and absolute power corrupts absolutely'. We cannot spare the criminals of the past who succumb to such temptation. Acton would hang such criminals, 'higher than Haman for reasons of quite obvious justice, still more, still higher for the sake of historical science'.[60]

Creighton responded as he always did to Acton's criticisms, with kindness and deference but this time also with subtle criticisms of his own. He admitted to being 'very grateful' for Acton's critique, that it was an act of 'true friendliness' and he was greatly encouraged for being criticized from such a high standard. 'Judged by it I have nothing to say except submit; "*efficaci do manus scientiae.*"' He believed that Acton conceived of 'history as an Architectonic for the writing of which a man needs the severest and largest of training'. He agreed that this ought to be the case but he had to 'admit that I fall far short of the equipment necessary for the task that I have undertaken'. He also claimed that 'I certainly agree with your principles of historical judgment' but was clearly more willing to accept different standards for different times in a way that Acton would not. He could not help but view the great statesmen of the past not with horror but with pity: 'who am I to condemn them?' Creighton asked rhetorically. 'Surely they knew not what they did.' He admitted that perhaps this was a foolish opinion and accepted Acton's general view in principle. He even hoped that someday Acton might write an article on the 'Ethics of History' for the *EHR*. 'I have no objection to find my place among the shocking examples.'[61]

Creighton was less gracious about Acton's criticisms when he discussed his review with others. He wrote quite candidly to sub-editor R. Lane Poole about the 'absurdity' of the *EHR* acting 'as a vehicle for making an onslaught on its editor'. He joked that he was 'tempted to add a note to the review, "The Editor is not responsible for the opinions expressed in the above article"'. It was rather amusing that an editor would invite and publish 'a savage onslaught on himself' but Creighton was upset that 'Acton does not clearly see what he has done'.[62] He wrote to Freeman that he could not possibly have engaged in the kind of judgements that Acton wanted to see. He was 'content to treat the actual facts and the traceable causes of them' while leaving 'theologians to do other work'. Moreover, he doubted that it would be possible to please Acton anyway. He 'thinks everybody a villain and judges everybody from the standard of an advanced liberal of today: he thinks that history is a branch of Ethics and ought to heighten men's conscientiousness'.[63] Clearly Creighton did not agree with Acton to the extent that he led Acton to believe.

The review was heavily revised before it appeared in print.[64] Acton tried to make his general criticism more clear while toning down some of the negative comments. The printed version does not read as if written by an enemy, though his general criticism stood: Creighton failed to live up to the true standards of scientific history where the Christian moral code must act as the historian's guide in truly impartial narratives. 'It is the office of historical science to maintain morality as the sole impartial criterion of men and things, and the only one on which honest minds be made to agree.'[65]

Acton may have believed that his method was more plain and simple than Creighton's, but as we will see (Chapter 7), he was largely paralyzed by his own method that taught the historian to be impartial and detached in order to let the past speak while also holding past historical actors and actions accountable to an inflexible Christian moral code. The *English Historical Review* gave him a venue to spread this message in his several reviews but he would never write a monograph-length study that would have better illustrated his method. It was not until Acton received the coveted Regius post, that he was able to implement more explicitly his vision for the practice of a truly scientific history. Such came via a suggestion from the Cambridge University Press Syndics in 1896 that Acton should direct a multi-volume project on the history of the world.[66]

Acton took up the challenge of engaging in this largely unprecedented project that he envisioned not as an encyclopaedia nor as a summary of what was already quite well known. He proposed a history of the modern world, a history of the West from the Renaissance to the present that would be divided into chapters written by the best-known specialists in their particular field. This was a grandiose undertaking but Acton believed he would be able to convince the larger historical community to take this on and what is more he believed that he could establish a method that would make such a project work. It would be a fitting monument to the historical knowledge produced in the nineteenth century, a product and symbol of history's scientific moment.

In a report provided to the Cambridge Syndics in 1897 Acton stated that the proposed project would include at least 254 chapters but would not exceed 260. He had already secured 96 writers who would write the majority of the chapters but still needed to assign 56 more. He admitted that 28 writers declined to participate either because they did not have the time or because of their age. A few others declined because the chapters suggested simply did not suit them. Furthermore, a handful of 'desirable contributors who had undertaken certain portions of the work have unfortunately died'. Acton was, however, less concerned about finding enough people; indeed, he felt he could produce the volumes with 60 competent men. What was most necessary, at this stage, was ensuring that the authors adhered to the same principles. With this in mind Acton proposed 'issuing an editorial circular to our contributors, laying down certain rules that may enable us to maintain something like unity of tone and treatment.'[67]

Acton felt it was entirely necessary to have specialists write on their particular subjects but this also necessitated having upwards of one hundred authors. This presented a real problem for Acton to overcome because he did not want the volumes to read as if they were penned by a hundred different voices. Indeed, he wanted it to appear as if penned by the historical community itself. This would be no easy feat but Acton felt as if his chosen authors could achieve it. This was, in fact, one of the goals he set for the project when he wrote a formal proposal to undertake the venture. The circular he sent to his authors included large extracts from that original proposal.

As he explained in his letter to contributors he wanted to produce the 'best history of modern times that the published or unpublished sources of information admit'. He believed that now was the best time to do this because 'nearly all the evidence that will ever appear is accessible now' – a statement that seems laughable a century later – and that there was a certain amount of unity within the historical profession to come together and present this knowledge. It is in this way that 'we approach the final stage in the conditions of historical learning'. In order for the volumes to be useful, however, he argued that the pieces needed to live up to the current standards of scientific history, standards that would allow for a certain unity despite the fact that the finished product would be a multi-authored venture. With this in mind total impartiality was necessary. 'Our scheme requires that nothing reveal the country, the religion, or the party to which the writers belong.' He impressed on the authors that 'any disclosure of personal views would lead to such confusion that all unity of design would disappear'.[68] He seems to have realized, however, that getting his authors to filter their narratives through a universal moral code expected too much and he never made this a requirement for the project.

The final extract from Acton's Report to the Syndics also spoke to the issue of impartiality and disinterest and it reads very much like his inaugural lecture at Cambridge.

> Contributors will understand that we are established, not under the Meridian of Greenwich, but in Long. 30° W.; that our Waterloo must be one that satisfies French and English, Germans and Dutch alike; that nobody can tell, without examining the list of authors, where the Bishop of Oxford laid down his pen, and whether Fairbairn or Gasquet, Liebermann or Harrison took it up.[69]

This was precisely Acton's attempt to take advantage of the *sensus communis* of the historical community, to take advantage of the hive of historical workers envisioned by Stubbs, and put them to work on a grand historical venture that was not the product of individual genius but of the diligent work of a community of scholars seeking to achieve the same end: a history from nowhere. Roland Barthes would refer to such an ideal as essentially a 'referential illusion', an attempt by historians to disguise their own highly subjective discourse as history itself. The historian is essentially '"absent[ing] himself" from his discourse' while substituting an '"objective" persona' and he reproduces not the reality of the past, which is in theory the goal, but merely a *realistic effect*.⁷⁰

For such a reality effect to be achieved (though Acton certainly would not have seen it in such terms), Acton needed well-known scholars to lend the project their names, despite the fact that the individual personas should not be readily apparent through a simple reading. Ironically, by entirely suppressing their individual subjectivities, historians were actually enhancing their professional authority as the only ones capable of presenting the past as it actually happened.⁷¹ There was no question as to whether or not the historians would sign their names to their individually authored chapters given that the whole practice of anonymity had been largely renounced by this time, a relic reserved for unoriginal and popular works that simply could not be trusted.⁷² Acton needed the authority of professional historians who could be trusted to suppress their individual personas and let the past speak for itself. One of the authorial names he approached as vital to the project was none other than Mandell Creighton.

Just as Creighton turned to Acton to write the introduction to the first issue of the *EHR*, it was to Creighton that Acton turned when he needed to find someone to pen the introduction to the *Cambridge Modern History*. Creighton had originally been asked and had agreed to write a substantial opening chapter for the volume on the Renaissance Papacy.⁷³ He had to withdraw from such a task after he was appointed Bishop of London and was unable to find the time necessary for the kind of specialized study Acton expected. Acton was convinced the loss of Creighton from the volume would be 'a serious blow to us, and especially to my position with the Press Syndicate. For my merit was that I got hold of the best men.' The press did expect Acton to write some sort of introduction that would state 'what our design is, what we mean to do, and not to do, and defining our notion of history, and what is possible in an impersonal work, written in these conditions'. Acton admitted that this was not a task he wanted to undertake and as he assumed it would be much less of an effort than a specialized study he asked Creighton if he would do so. 'I want to ask you to write the Introduction, and inaugurate the undertaking', he wrote to Creighton. 'You are so preeminently the man for it, by your present dignity as well as by your former connection with history at Cambridge and with the Review.'⁷⁴ It seems that Acton had developed the same editorial caution as had Creighton ten years before. In this case, however, Creighton agreed to write the introduction.

It may seem odd that Acton would ask Creighton to pen the editorial introduction given the 'yawning difference' that Acton witnessed between their historical methods. However, it is clear that for this project that gap did not exist. Acton needed his authors to suppress their subjectivities and perhaps this more mechanical approach was deemed most important beyond ensuring that a universal moral code be followed. Creighton's introduction showed that he understood the project and its methodology only too well.

Creighton sought to explain the method that was behind this grand project and the 'exceeding difficulty of writing a history of modern times on any consecutive plan'.[75] He explained first just how difficult it was to decide when modern history began, a debate that had been somewhat divisive at both Cambridge and Oxford over the past century. But that largely endless debate had to be ignored in this instance and arbitrary divisions implemented if only because the narrative had to begin somewhere. It was decided that the best that could be done was to arbitrarily begin where 'mankind' reaches 'the stage of civilization which is in its broad outlines familiar to us, during the period in which the problems that still occupy us came into conscious recognition, and were dealt with in ways intelligible to us as resembling our own'.[76]

Creighton explained that there was an even more difficult element to the project and that simply related to just how the story of modern history would be told given the fact that it will be told by multiple authors. Creighton implored the reader to comprehend the immensity of the task at hand and how it would not be possible in the age of science for a single author to undertake it. Such would simply not live up to the universally accepted standards of scientific history. 'It is no longer possible for the historian of modern times to content himself with a picturesque presentation of outward events.'[77] The Romantic or picturesque historian is largely unconcerned with building his or her historical narrative on an inductive basis but does so on the basis of his own genius. It has long been established that for history to be written adequately, however, 'accurate facts are needed, – not opinions, however plausible, which are unsustained by facts'.

He explained that the scientific view had become so dominant in part because of the immense expansion of archival sources that became available throughout the last century, a process that necessarily antiquated even the most recent historical studies. The only way to keep up with this massive accumulation of knowledge is to become a specialist in a particular subject matter and engage in a thorough examination of the available sources. And such a specialist must continue to study the new monographs and new archival materials that inevitably appear in order to stay abreast of recent trends and facts as 'new knowledge is always flowing in'. 'Modern history in this resembles the chief branches of Natural Science; before the results of the last experiments can be tabulated and arranged in their relation to the whole knowledge of the subject, new experiments have been commenced which promise to carry the process still further.' And in some senses, Creighton explained, history is more difficult in this regard given that its subject matter changes continually while science's 'object of research is fixed and stable'.[78]

It was for this reason that the project was completed by specialists of particular historical subjects and the narratives resulted from these individual studies. The narratives 'are not derived from previous conceptions of necessary relations between what he has studied and what went before or after; they are formed directly from the results of the [historians'] labour'.[79] It is in this way that 'the subject-matter [will] supply its own unifying principle' and 'the age [will] be presented as speaking for itself'.[80] Acton could not have said it better himself.

Before the first volume was published a preface was also added to explain that the project was planned by Lord Acton. The editors explained that much of the work followed Acton's initial vision from the division of the volumes and the chapters through to the authors chosen. The editors 'adhered scrupulously to the spirit of his design, and in more than one passage we have made use of his words'. They went on to describe the great 'cooperative principle' envisioned by Acton, one that required an immense 'division of labour' and a commonality of method.[81] The preface became necessary when Acton died just before the first volume was completed and sent to the printers. It is somewhat ironic that even though Acton did not want to appear in the first issue his words were borrowed by the editors in order to explain the purpose and method behind the volume, perhaps finally achieving *le désintéressement des morts*.

6 BROAD SHADOWS AND LITTLE HISTORIES

I am beginning to think that this General Reader is after all, not quite such a fool...

Edward A. Freeman to James Bryce, 24 August 1873

The combination of readableness and research is so difficult as to be almost impossible...

Mandell Creighton to W. E. Gladstone, 15 February 1887

The establishment of the *English Historical Review*, the *Cambridge Modern History*, and even a few well-researched scholarly monographs held out the hope and promise that an inductive science of history would become the rule rather than the exception, that histories from nowhere would essentially be everywhere. The reality of history publishing was far from this ideal, however. Scientific historians may have wanted to present the past in all its reality, but publishing such work was the only way to communicate the historian's findings and this left historians largely at the mercy of publishers who were themselves dependent on a readership not necessarily interested in scientific standards. For historians who were supposed to let the past speak for itself, much of their work relied on just the style they condemned in others and was directed specifically at the readers who would have had little interest in reading good scientific history. As much as the Rankean historians expressed a desire to create a normative world where their work would be read and judged by peers alone, they were still very much public figures who wanted their views concerning history to be both widely disseminated and appreciated.

As Stefan Collini has shown, historians – as well as men of science – in the late Victorian period often had to adopt different identities depending on the particular audience they sought to address.[1] Indeed, many were forced to embrace dual personas: the pedantic upholder of scientific standards for the world of peers; and the public intellectual disseminating his critically researched scholarship to a 'general' audience. These identities were not necessarily at odds but often the latter had to transcend the very boundaries erected by the former.² Several historians of the scientific persuasion sought to publish just that scholarly monograph intended for peers while placating their publisher and also supplementing their income by printing something directed at a more popular readership. A process seemed to develop whereby historians who were able to prove their scientific persuasion could also be trusted to speak to a wider audience without corrupting the public's historical sensibilities. Indeed, they could even be trusted to subtly impart the new scientific standards to the broader audience while writing in a style that would normally be scorned. One had to be an accepted member of the scientific community of historians in order to do this well, however, or else suffer the wrath of the discipline's boundary work.

Edward A. Freeman, as we have seen (Chapter 4), was outspoken in his criticism of a form of historical writing that did not live up to his own normative scientific standards, and yet he also felt free to write for popular audiences, though such work was always secondary to his more serious monograph-length studies. His first major work of history was a *History of Federal Government*, with the first and only volume appearing in 1863 on Greece and Italy. Freeman was forced to directly confront the fact that such scholarly work was not in demand as Alexander Macmillan would only agree to print the book should Freeman cover half of any debt accrued by the publisher.³ The sales were disappointing to say the least and Freeman hoped he could pay his half of the debt in instalments otherwise he would need to take out a loan to cover the balance of £130.⁴ Freeman believed that the book's failure to sell was due, not to the subject matter or his rather repetitive and dry writing style, but rather to Macmillan's lack of advertising, about which Freeman complained bitterly.⁵

By November 1865, Freeman suddenly decided to put on hold the later planned volumes of his *History of Federal Government* when he heard that Goldwin Smith had decided to resign his Regius professorship at Oxford and Freeman realized that he needed to get something else published as soon as possible.⁶ He had several years earlier been spurred into writing a history of England for his children by Kingsley who had been supposedly writing a history of England for boys.⁷ Macmillan was clearly interested in the study and pushed Freeman to complete it and submit it for publication.⁸ Freeman renewed his interest in completing his 'little Early History of England' once he learned of Smith's intentions, a book he could write much faster and more easily than another volume of *Federal Government*. What was more, Freeman proposed that the book would not only teach children about the early history of England but that it should also introduce children to the methods of scientific history. He was convinced that if children could be taught such principles at an early age, they could avoid being corrupted by popular historians later on. He claimed that the main 'object of the book' would be to give children accurate and scientific views of history from the very first, to teach them to call things by their right names, to distinguish history from legend, to know what the sources of history are, and to distinguish the different values of different writers. I can only say that, with my own children it thoroughly succeeds. To many people I dare say it would seem hard, what they would call over-learned. The truth is that the scientific way of doing anything is puzzling to those who have learned some other way, and who are called on to *unlearn*. It is not puzzling to a child who has learned nothing; quite the contrary, because the scientific way is really the easiest because the clearest. This is true where I can judge, namely of history and language, I dare say it is equally true of all other subjects.

Freeman was clearly excited by the prospect of teaching children the right and proper way of learning history before having to convince them later on to 'unlearn' what they had wrongly been taught.[9]

Freeman's enthusiasm for the little book declined quite rapidly after he talked to both Smith and James Bryce about it. They both seemed 'to think anything like a child's book might do me more harm than good', he wrote to Macmillan, 'as an adversary might take advantage of it to make a child's book my measure'.[10] Indeed, Freeman was worried that someone might criticize him in the same way he denounced Kingsley's published lectures as being the product of a children's author, good enough perhaps for a 'land-baby' but not for undergraduate students (see Chapter 3).[11] Freeman therefore proposed writing at the same time a more specialized study that would overlap with the subject matter of the little history. This would be a 'History of the Norman Conquest', something that he could write quite quickly as it was a subject he had been working on 'for twenty years'. He had actually submitted a study on the topic in 1846 for the chancellor's English prize essay while he was on a fellowship at Balliol College. The essay did not win the coveted prize, but Freeman felt that now was the perfect time to revisit the study and have it published at roughly the same time as his book for children thereby at once filling out his scholarly publishing record while also placating Macmillan who was keener on the history for children.[12] Indeed, Macmillan would eventually convince Freeman to publish his more specialized study of the Norman Conquest study with the Clarendon Press Series at Oxford.[13] Freeman was concerned that Oxford books get ignored because its press 'used to take every possible means to hinder the world for knowing of their existence'. He hoped that Macmillan had 'taught them better' than their reputation suggested.[14]

Freeman would of course lose the Oxford position to Stubbs, though he would do his best to irritate the Clarendon Press to no end by trying to get his first volume published in time for consideration, and it would appear as *History of the Norman Conquest* in 1867 (see Chapter 2). Four more volumes would be printed in fairly rapid order with the final volume being published in 1872. The first volume was fairly well received though the research was entirely based on secondary and printed primary sources. This would be a general theme of Freeman's serious scholarship. He often made great use of the Rolls Series and other printed sources but he rarely saw the inside of an archive.

Two years after Freeman's identity as a scholarly historian had largely been secured by his *Norman Conquest*, Macmillan published Freeman's *Old English History for Children* (1869), a work that was deemed to be much more successful than any of Freeman's previous ventures. Quite in contrast to Freeman's *History of Federal Government*, Macmillan agreed to take full responsibility for any debt the book might incur, accepting the sole risk and expense of the first and any future editions of the book. This was because the book was, unlike Freeman's first, expected to make money. Freeman was even advanced £100 for the royalties of a first printing of 4,000 copies.[15]

Interestingly, the book was expected to sell despite the fact that Freeman did not deviate from his original plan of teaching English children the principles of scientific history. In the preface, Freeman made this clear, arguing that the work was somewhat of an experiment to show that 'clear, accurate, and scientific views of history, or indeed of any subject, may be easily given to children from the very first'. He observed that children tend to understand 'the more rigidly accurate and scientific statement'. 'The difficulty', he said, 'does not lie with the child, who has simply to learn, but with the teacher, who often has to unlearn'. Much like the general reading public, the teacher, 'has often been used to a confused and unscientific way of using words'.[16] Upon reading the book, J. R. Green admitted that he greatly admired it and assured Freeman that 'it is certain to be popular, and … do an immense deal of good'. Green particularly appreciated the 'introduction of … children into the whole criticism of authorities, etc. *This* constitutes the real originality and value of the book.'[17]

Freeman also claimed that he received 'all manners of letters about the Little History. 'Some want to strike out the words "for children"', suggesting, of course, that the lessons one learned about history throughout were ones appropriate for adults as well.[18] One letter particularly caught Freeman's attention, from Bettie, aged ten:

> Dear Mr. Freeman,
> I am a little girl of ten years old, and I have read your *History of England*, which I like better than any book of history I have ever read. I am writing my life, and am going to write a description of your History of England, and so I should like very much to have your photograph to put it in. … I like Harold much better than William, and I do wish you would write a History of England after the Conquest, as you said you would at the end of your book. It would be so much clearer, truer, and more interesting than what other people write.[19]

Freeman was brought into contact with Bettie through his friend Lord Carlingford (C. P. Fortescue), a friend of Bettie's family. In Carlingford's letter, that included Bettie's, he claimed that Freeman 'had inspired this Betty [*sic*] with a deep distrust of other historians, and ... she often says, "I wonder what Mr. Freeman would say."'[20] This deference to the proper historical authority was exactly the effect Freeman hoped his book would achieve. Freeman sent the photograph and responded: 'I am delighted to find that my History is doing good just where I wanted it to do good. So go on and prosper, and when you have written your life, let me see it.'[21] Bettie in turn responded: 'I am afraid I cannot show you my life. It is only about foolish things you would not care for, and I only show it to mama.'[22] Bettie learned that some things should be left to professionals.

Leslie Howsam argues that by publishing *Old English History for Children* Freeman 'shows a fine disregard of the boundary constructed by his peers, between the apprentice historians at Oxford and Cambridge sitting at the feet of Stubbs or Acton and painfully learning to practice history as a science, and the general reading public'.[23] But this was a boundary Freeman was only too conscious of transgressing. He was sure to complete his study of the Norman Conquest first and he liked to downplay his little history to his peers, often suggesting that it was a trifling book he was merely writing for his own children, even after Macmillan had already agreed to publish it.[24] Freeman's *Old English History for Children* surprised many but it also set a precedent for professional historians looking to sell copies of their works and make money doing so. Freeman could write a history for children because he could be trusted in his account of the past, no matter how sweeping his generalizations. Historians certainly took note of this, as did publishers who were now on the lookout for competent historian-authors capable of disseminating their work to a wider audience.[25] Macmillan did not have to look beyond Freeman's close circle of friends to find another competent historian-author who also just so happened to be highly impressed with Freeman's more popular study.

John Richard Green (1837–83) was one of the primary members of the community of scientific historians who also spoke to a wider audience. He grew up in a strict Anglican and Tory household, though he rebelled against his upbringing by adopting more liberal views. He received a scholarship to attend Jesus College in Oxford in 1856 and was deeply disillusioned by the system of patronage learning, much like Kingsley, and received a mere pass degree in 1859. He would begin his historical work by writing a series of articles on eighteenth-century Oxford life for the *Oxford Chronicle*. He also took deacon's orders in 1860 and the following year was ordained as a priest spending the next ten years working in the most impoverished parishes of London. This work was rewarding for Green but it contributed to worsening the tubercular condition he had contracted as a young child. He eventually would end up as the librarian of Lambeth Palace from 1869 to 1877, a position much better suited to his health.

Green had first met Freeman as a student at Magdalen College School and the two became reacquainted when Green presented a paper before the Somerset Archaeological Society in 1862. He somehow managed to endear himself to Freeman and soon after to Stubbs. Both believed that he had the makings of an excellent historical scholar (though, as we will see below, Stubbs was not immediately convinced of Green's capabilities). His librarianship at Lambeth allowed him to pursue his historical interests with more focus and Green wrote over 160 review articles for the *Saturday Review* between the years 1867 and 1873, and was along with Freeman part of the reason why the journal became such a strong proponent of scientific standards in historical writing.[26]

Green had a much more radical worldview than did his conservative friends. He was an outspoken Christian socialist and he was not ambivalent, like many other historians, about Darwinism. It was largely through his recollection of the famous debate between Huxley and Wilberforce that we learn that Wilberforce attempted to demolish the Darwinian hypothesis but it was Huxley who was victorious, defending Darwin by giving 'his lordship such a smashing as he may meditate on with profit over his port at Cuddesdon'.[27] Stubbs cringed whenever he had to suffer through one of his friend's rather loud and radical rhetorical flourishes.[28]

Furthermore, Stubbs was initially torn about Green's capabilities as an historical worker. He was actually quite cool to the suggestion that Green replace him as editor of the Rolls Series once he, Stubbs, was appointed Regius Professor of Modern History, believing that Green was 'far too unmethodological and in fact careless' and he also felt that Green was 'a little too much into theories', not a minor criticism coming from Stubbs.[29] Freeman, however, was convinced of 'Johnny' Green's capabilities. 'I am very fond of him indeed', Freeman explained to Bryce, 'and as for his ability and research I can truly say that the difference between him and Stubbs is simply a difference of age'.[30] This was high praise indeed.

Stubbs was eventually persuaded by Freeman to appreciate Green's particular historical abilities and the two of them often 'tooted' Johnny Green's horn to friends and publishers.[31] Green certainly appreciated this praise but he often failed to return it, particularly when reviewing his friends' works. He often criticized Freeman for writing old-fashioned 'drum-and-trumpet' history, the history of kings and the aristocracy, of wars and high political intrigues. Green's review of Freeman's *Norman Conquest* certainly did not appear as if written by a friend.

> There is evidently a powerful attraction for Mr. Freeman in the outer aspects of wars and policy which throughout tends to lead him away from the examination of those deeper questions which lie beneath them. His book is not, we think, sufficiently

penetrated with the conviction of the superiority of man in himself to all the outer circumstances that surround him. We are, or course, far from classing the *History of the Norman Conquest* with the mere 'drum and trumpet' which Dr. Shirley so pungently denounced, but throughout there is too much of wars and witenagemots, and too little of life, the tendencies, the sentiments of the people.[32]

Freeman could hardly believe that Green would be so critical of his good friend's work. He wrote to Green that he must understand that his criticisms would be quite damaging. He tried to explain that when reviewing a friend's work, criticism is fine and can even be productive but the review as a whole should help promote the study in a positive way rather than do damage to it. 'My review of Stubbs will do him good', he explained to Green, while 'yours of me will do me harm, tho' I am sure that nothing is farther from your thoughts'.[33] This was a fairly constant debate Freeman would have with Green who never seemed quite willing to play the game by Freeman's rules. Freeman felt that Green was simply condemning him for not writing the kind of history that Green wrote, that is, an early form of social history.

Despite Green's published criticisms of his friend's work, Freeman seemed to think this a minor irritation rather than anything worth ending a friendship over. What is more, Green was extremely outspoken in his praise for Freeman's *Old English History* and it is clear the work had a great influence on his own writing, though it took him a while to find an appropriate subject matter worth studying and writing about at great length. He had attempted to write a specialized history of the Angevin Kings, but was overwhelmed by the project. Freeman's *Old English History for Children* pushed him in another direction altogether, as did his growing friendship with Macmillan. In 1869 he wrote to Freeman:

> I have offered Macmillan to write a *Short History of the English People*, 600 pp. Octavo, which might serve as an introduction to better things if I lived, and might stand for some work done if I didn't. He has taken it, given me £350 down and £100 if 2000 copies sell in six months after publication. He seems delighted with the sale of your little book – 1200 gone already.[34]

Freeman's little book had done quite well and Green also wanted to write something that more than a few specialists would read. This was in part because he was not healthy and was unsure how much longer he would live. He wanted to publish something that he would be remembered for.

When he approached Macmillan about the proposed book, he was very clear about the subject matter and also the particular audience the book would target. He said that he wanted to write a history of the 'English People', one where 'war and vulgar events would occupy a far more subordinate position than they generally do'. The focus of Green's book would instead be 'directed to the growth, political, social, religious, intellectual [world] of the people itself'. 'The style of such a book', Green explained, would be 'more picturesque ... than if it were on a larger scale'. He wanted 'especially [to] avoid cramming' the book with intricate 'details'. In this way, the book would hopefully serve as a 'school-manual for the higher forms', perhaps acting 'as a handbook for the universities'. He believed that such a book might 'supply a real want in our literature – that of a book in which the real lines of our history should be fixed with precision and which might have as an introduction to a more detailed study'.[35] Macmillan clearly tended to agree with Green, given the contract the two of them signed and the sizable advance he gave to Green against the book's expected sales.

Both Freeman and Stubbs were taken slightly aback by Green's new project. Freeman wanted Green to continue with his more specialized study of the Angevin Kings and establish himself as a professional scholar of history before publishing a more popular study. At the very least, he hoped he would write them at the same time.[36] Stubbs quite simply believed that such a popular study of English history would be unnecessary. 'For a popular History such as he contemplates', Stubbs wrote to Freeman, 'surely, Charles Knight and the Pictorial people have done what is necessary and possible from existing materials'.[37] As far as Stubbs was concerned, Green was merely reproducing something similar to Charles Knight's *Popular History of England* (1856–62), an eight-volume study written for mass consumption reliant on a host of illustrations to help present the past in a much more palatable format, though Knight would claim that the many wood-engravings were selected 'not as mere embellishments but as illustrations of the text'.[38] Stubbs thought Green's project 'absurd' at least the way it was presently formulated. However, Stubbs told Freeman not to worry too much about Green's current false step 'as the dear child never does anything he says he will, I think we may set our minds at rest'.[39]

Green would prove Stubbs wrong on both counts, though it appeared, at least for a while, that this would not be the case. When Green completed and submitted the first chapter it appeared as if he and Macmillan had a very different understanding about how the book would look, despite Green's fairly clear proposal. Macmillan did not like the first chapter. His editor, George Grove, sent Green comments on the proofs of the chapter and complained about the general readability of the work. Grove suggested that the chapter be entirely rewritten in a way made interesting for the general reader. Green had written a chapter so detailed and filled with dry factual data that only specialized historians would have found it interesting. Writing to Freeman, Green admitted that perhaps Grove was right, that the book was unfit for the general reader. This was a problem because Green had agreed to write such a work. 'About Macmillan I have a very strong feeling of honour – I offered to write a book for "general readers," and I can't hold him to his engagement if the book is – as it is – unfit for them.' What is most interesting is that Green practically apologized to Freeman for agreeing to rewrite the chapter along the lines set out by Macmillan: 'Please don't think me despondent, – I want to be cool and fair, – and I am resolved to write *something*; that is to say, if Macmillan agrees as I think he will I might still try to rewrite this chapter in narrative form, leaving out 50 per cent of the matter I have packed so tight, and chattering more diffusely over the rest.' Green suggested to Freeman the possibility of scrapping the whole project and revisiting his work on the Angevin Kings, a suggestion which Freeman thought Green should strongly consider.[40]

Green kept at his little history, however, and in 1874 published a remarkably successful *Short History of the English People* that did much better than either Green or Macmillan could have planned or even hoped for. Macmillan even destroyed their original contract in favour of a royalty agreement when the little book sold 8,000 copies within the first few months, an agreement that would pay Green 18 pence for each copy sold.[41] The book would sell 32,000 copies in its first year alone and over 500,000 throughout the rest of the century. Only Macaulay's *History of England* could claim this kind of popularity with England's reading public.[42] The book was for the most part favourably reviewed. The *Saturday Review* found it to be filled by a 'wealth of material, of learning, thought, and fancy' striking the reader with a 'freshness and originality.'[43] However, heavy criticism did appear in a review in *Fraser's Magazine*, which pointed to several factual errors.[44] This review did not faze Macmillan who, by 27 March 1876, could argue that the 35,000 copies sold 'is the best proof' that the review's criticisms 'were of slight significance'.[45] Indeed, for Macmillan, copies sold spoke loudest as to the value of the work.

Freeman certainly would not have agreed with such an assessment but he was one of the most outspoken supporters of *Short History*. He gave Green comments and much encouragement throughout the writing of the book once it became clear that Green had abandoned his Angevin study. He believed that Green was on to 'something quite new' even if he 'did not above half understand' and he encouraged Green to by 'all means do it your own way, and let it take a chance. A great deal of it is ... so powerful and striking that it must succeed.' With that said, he believed that Green had 'largely sacrificed the real stuff to your power of brilliant *talkee-talkee*, but I know that a great many people will think otherwise; so go on and prosper in your own fashion'.[46] When the book was finally published, Freeman 'rejoice[d] greatly' at the book's success.[47] While Freeman believed that Green's earlier historical studies were much better,[48] and he had significant points of concern with the published version, he reviewed the work extremely well in the *Pall Mall Gazette*.[49] Freeman hoped the friendly review would act as an example Green would follow in future reviews of Freeman's work. Freeman stated that he was certainly 'true and fair' but had it 'not been Johnny Green who had written the book, I might have enlarged more fully' on some of the minor criticisms. Freeman hoped that Green would now understand 'what I have complained of when you have been reviewing me'. 'You see that I can see merit in a book what is utterly different from anything that I should have written.' History books, whether written by 'I, Stubbs, Bryce, anybody else, will each of us have his own way' and he hoped Green would now understand 'that each will do it best by doing it in his own way, and that he should be judged by that standard. Remember this another time.'[50] It would be up to the *Quarterly Review* to denounce the way in which Green's style trumped accuracy while lamenting his failure to focus on the monarchs and military history of England.[51]

Despite the fact that Green continued to refuse to write the friendly review, his book was well received in part because of his perceived position within the nascent historical community. What is more, he took advantage of that position. Anyone who may have been unaware of Green's membership in the community of scientific historians only had to read the preface to *Short History* where Green mentioned the 'invaluable Constitutional History' of Professor Stubbs as well as the constant 'counsel and criticism' of 'my dear friend E. A. Freeman'.[52] Green connected himself to the established authorities on English medieval history, which led the reader to conclude that like Freeman and Stubbs before him, Green, too, was a trusted authority.

While Freeman was not entirely uncritical of Green's very popular presentation of English history, arguing that 'it is full of his strengths as well as his weaknesses',[53] it was still able to convince him that perhaps writing for the general reader was not a complete waste of time after all. In a letter to Bryce where Freeman sought to explain the plan of Green's work-in-progress, he admitted that 'I am beginning to think that this General Reader is after all, not quite such a fool as I [had originally] ... thought [of] him'. He had already written a volume for children, but Green's project convinced him to write something more popular for an older reader. As he was beginning to understand, '[t]he way to keep [the general reader] from being a Fool', he wrote to Bryce, 'is to treat him as if he were not one'. He believed that 'Johnny's book would [do] just' that and perhaps he could take a similar approach while promoting the same method that worked so well for children to young adults and students as well.

Instead of just writing another popular volume, however, Macmillan convinced Freeman to edit a series of historical works, a 'Historical Course for Schools' that would produce a series of works, each devoted to a different European country. Freeman decided to pen the introductory volume on the 'History of Europe' as a shilling book and he set about finding the appropriate authors for volumes on England, France, Greece and others.[54] Interestingly, Freeman specifically sought out women authors to write the volumes in the series, though in the end was only able to secure three and had to find male authors for the remaining works.[55]

It is not entirely clear why Freeman decided that 'she-bodies', as he called them, were the ideal candidates to write for the 'Historical Course for Schools' series. It is somewhat ironic that as women were largely excluded from the formalized university route to academic professions such as history, they were becoming more central to the education of children. For all the rhetoric concerning the masculinization of knowledge during this period, women were considered excellent educators and their labour was essential in the education of the nation, particularly following the Education Act of 1870, which made elementary school mandatory.[56] While it was believed women could make no great discovery on their own, it was also believed that women could condense and disseminate knowledge in a nurturing manner for children and the uneducated. Indeed, when it came to teaching, women could appropriate their 'special qualities', such as their nurturing nature, to justify their entrance into a profession that was desperately in need of elementary educators.[57] Women were particularly central in the popularization of contemporary scientific knowledge,[58] and Freeman must have believed that they would also be useful in disseminating historical knowledge if guided by his own professional acumen.

Freeman was initially hopeful that his twenty-year-old daughter would write the volume on Greece (which was not published) but he was also aware that she had friends who had sufficient knowledge to be put to work on specific subjects, particularly Edith Thompson, who would write the work on England, the most popular volume of the series.[59] Upon hearing about the authors for Freeman's series, Green was fascinated and began referring to them as Freeman's 'historic harem'.[60]

Once word got out about Freeman's selection of authors, he began to receive unwanted advice. Green discussed Freeman's scheme for the series with Bryce and relayed Bryce's message that Freeman should stick to established authorities. Bryce 'said one noteworthy thing', Green wrote to Freeman, 'that these little things must be done by big people – that they are the most difficult things of all to do, and that till big people can find time to do them they had better wait. We have enough bad work already'.[61] Freeman would have readily agreed that there were enough bad popular works of history on the market. Indeed, about fifteen years early he had denounced the 'torrent of Lives, Memoirs, Sketches, &c., &c.' that tended to be written by women authors as 'the greatest possible hindrance to historical knowledge',[62] and Freeman was certainly not going to attach his name to such poor historical scholarship.

Perhaps Freeman chose women as his authors because he felt they could relate to the audience targeted by the works: those learning history. As editor of the series, Freeman could take each author under his wing and while teaching the authors about the critical standards of a professional history, that process could somehow be mimicked in the writing of the work; meanwhile the works, because of their feminine touches, might tap into that popular readership that female authors seemed so easily to reach. If this was the case, there is also some evidence to suggest that Freeman was more than successful in mentoring Thompson in particular to the necessary dryness of professional scholarship. After reading through a draft of Thompson's volume, Green could not quite believe just how boring it was. '[A] capital piece of work done by a clever woman', Green wrote, 'and as dull as an old almanack! I daresay governesses will find it "useful," but it will set every child against a study so absolutely without human interest.'[63] Again, six months later, Green sent comments on the general series as well as the work on Scotland which echo his earlier critique:

> As to the general plan I again deplore – as I did in the other case – the *entire exclusion* of all stories, anecdotes, or anything which by any possibility can enliven the tale. As it stands the book is *utterly unreadable*. Of course this is a matter which rests wholly with you, but I do hope you will consider whether absolute dryness and unreadableness is a *sine qua non* in educational books. The style, too, is terrible. ... Did you issue instructions to your Harem strictly forbidding the Beautiful and the Interesting?'[64]

Green also told Macmillan that what was truly necessary was a 'suppression of useless details and condensation of dull and unmeaning periods'.[65] From Green's fairly bored reaction to the series while it was still a work in progress, one would have to conclude that Freeman did not embrace the feminine touches of his female authors.

As Howsam has shown, however, Freeman was most likely interested in securing unknown female authors because he could control them in a way he could not control members of the 'he-flock', a term he used to describe the members of the male historical community.[66] In a letter to Macmillan, Freeman responded to the suggestion that he approach A. W. Ward to write the work on Greece. 'My difficulty in asking' Ward, Freeman wrote, 'is the same as it would have been in asking Bryce or Johnny [Green]. If I am to have the direction and responsibility of the whole series, the other writers must be people who will knock under to me, people whom I can decently ask to knock under to me.' Indeed, Freeman did not want independent thinkers but rather authors he could place entirely under his historical authority. 'Now I can't ask this of a man like Ward. If he differs from me on any point, he has a good a right to his opinion as he has to mine.'[67] Freeman wanted to be in absolute control of the series, controlling not only its general shape, but the entire content as well. He did not want to be put in a position where he might publish a statement in the series that he might not personally agree with. Freeman was only too aware that any member of the 'he-flock' would likely, in his own words, do the manly thing and have the courage to hold 'his own judgment, if reason and conscience bid him' even if it was 'against his own friends, against his own side'.[68] From this perspective, Green's labelling of the series' authors as Freeman's historic harem hits the mark.

It was Freeman's duty to ensure that the truth of the past was conveyed and he expected his 'she-flock' to 'knock under' to his historical authority. Freeman seems to have employed his female authors in much the same way that American astronomer Edward Pickering employed his more popular 'harem' of female employees between the years 1877 and 1919 to engage in non-observational work that really just contributed to their employer's scientific work, rather than to their own.[69] While Freeman was only the series' editor, he made it clear that the historical observing was up to him. He may not have written the volumes, but he expected the works to fall in line with his own belief in what truly happened. He was actually quite disappointed when he had to approach both Green and Bryce to write works for the series.[70] Freeman would eventually fire Green as one of his authors, as Green refused to write the volume to Freeman's specifications. 'I think it would be better that you should not do it', Freeman wrote to Green. 'Your ideas and mine as to what it should be like differ so utterly.'[71]

Green would eventually edit his own popular series of history books for Macmillan and he was set about getting the big names that Freeman sought to exclude. One historian he convinced to write for his little series was the future Dixie Professor at Cambridge, Mandell Creighton. His *Primer of Roman History* for Green's series would be published in 1875 and sold extremely well. For all Creighton's insistence on producing specialized scientific history, the little book showed that he could write history for a broader audience if he set his mind to it. 'In its own way', argued Creighton's wife, Louise, 'it is one of the best things he ever did'. She also claimed that Lord Acton 'considered it the most masterly thing he had read for a long time'.[72]

Creighton's success with Green's series of historical primers led Creighton to edit his own series for the Rivington publishers entitled 'Historical Biographies', a project he undertook in 1874. Creighton would write the volume on Simon de Montfort. His long dealings with Charles Longman also began at this period as Creighton agreed to write a primer on the Age of Elizabeth for a series Longman was producing called 'Epochs of English History'. Creighton would also edit that series writing the volume the *Age of Elizabeth*, the introductory volume, *The Shilling History of England*, as well as a volume on the *Tudors and the Reformation*.[73] Like Freeman, Creighton employed a few female writers for the series, including his wife Louise Creighton and Samuel Gardiner's wife Bertha Meriton Cordery Gardiner. Louise Creighton would become an excellent historian in her own right, publishing several studies on periods of English history that relied quite heavily on primary sources.[74] Green's wife, Alice Stopford Green, would also become a well-known historian, not only publishing a revised and corrected version of Green's *Short History of the English People* but also publishing her own popular social history of the fifteenth century and a critical analysis of the gender of historical knowledge.[75]

If Creighton, Green and Freeman were more than willing to write educational, and, in the case of Green, outright popular studies for broader readerships than was perhaps suggested by much of the rhetoric that underpinned scientific history, the other major proponent of the scientific method of history, J. R. Seeley, was a more reluctant convert. We've already seen how his anonymously published *Ecce Homo* caused a sensation throughout 1865–6 (see Chapter 4), but the reception of the book led Seeley to be more wary about letting Macmillan have his way in marketing his future work. Seeley felt that he was forced to learn a hard lesson with *Ecce Homo* and he was not about to endure the pain of being harassed and misunderstood by easily confused, excitable and angered common readers. Indeed, as he explained to Macmillan, he was alarmed by the success of *Ecce Homo* and would have adopted an approach to exclude those readers unable to truly understand the message if only he knew how.[76] But, of course, he did know how to do so with any future work – by targeting a more scholarly reader not so easily excited by the book's message.

When Seeley finally did write the long-promised sequel to *Ecce Homo*, he refused to direct the book's message to the common reader. *Natural Religion* would be serialized in *Macmillan's Magazine* between 1875 and 1877 before being published as a book in 1882, and Macmillan hoped doing so would increase the interest in the book. However, when Macmillan began receiving parts of the study, he knew that it would not find a broad audience. Indeed, *Natural Religion* was almost unreadable and would likely be a failure. Seeley, again, wanted to publish the work anonymously but Macmillan felt that the book would be much more successful should Seeley sign his name as author. He was more widely known by this time as Regius Professor of Modern History at Cambridge and his name would have generated interest in the study even if the writing style was unattractive and might in the end dissuade readers. Seeley refused.

> It may seem to you whimsical that I should choose to run the risk of such a failure when I have a ready means of securing attention. It *would* be very whimsical if the question were of a poem or a novel, but on more burning subjects popularity brings much more pain than pleasure, and besides pain perplexity and anxiety. The truth is I had much rather fail as you anticipate than succeed in the other way.[77]

The other way to success was far too painful for Seeley to relive.

Seeley therefore sought to publish *Natural Religion* both anonymously and specifically for a much smaller audience than was targeted with *Ecce Homo*. As a result, the book was a commercial failure. Sometime in the 1890s, Macmillan approached Seeley about putting out a cheap edition of *Natural Religion*, and this time in Seeley's name. Macmillan was clearly hoping to get something out of a book that had come nowhere close to living up to its billing as the sequel to *Ecce Homo*. Seeley refused to budge. 'I am sorry to feel obliged to keep back my name, particularly as you say it would help you to publish it.... To write on these burning questions [is upsetting to] me more than others can easily realize.'[78] Even after Seeley died, Macmillan's successors were still trying to put out a version of *Natural Religion* worth buying. Even though Seeley's widow Mary could have used the royalties a new edition would have brought in she proved just as stubborn as Seeley on the issue. A cheap edition would not be possible, Mary claimed. *Natural Religion* 'was written for a certain class i.e. *educated* people who have a difficulty in accepting revealed religion, *not* for the general public'.[79] Macmillan knew too well that there would not be a wide enough scholarly audience to justify the publication of works specifically directed at them.

It is likely in part because of the failure of *Natural Religion* that Seeley allowed himself to be convinced to publish a series of his Cambridge lectures that would attract a wider audience and help Macmillan cover some of his losses.[80] Seeley was hesitant about publishing the lectures and he became quite nervous at Macmillan's suggestion to send them to the printers. The lectures, Seeley explained, were 'written in great distraction and without sufficient plan'. However, Macmillan could see the popular appeal of such lectures, where the general theme of the role of empire in the development of English history was apparent.[81] The lectures would appear as the *Expansion of England* (1883) and would become a huge success, appearing at the precise moment when arguments for and against a new imperial endeavour were being strongly debated. Seeley seemed to give voice to those who saw the British Empire as an almost natural extension of English power.

Initially, however, the lectures were not primarily even about the British Empire. Seeley originally wrote them as discussions of history's scientific methodology and Macmillan did a wonderful job suggesting editorial changes that foregrounded the discussions of empire while suppressing some of the specific historiographical points. Knowing readers will see that much of the discussion of empire is written within the context of a historiographical debate about the proper way to write a scientific study of England's history. Seeley was insistent that the narrative of England's history had to be told within the context of empire and he believed that historians had largely failed in making empire a central element of the story. Indeed, he argued that it was because of the previously flawed work of historians, and not politicians, that the British Empire seems to have been constructed in an 'absence of mind', a very different argument than the one that is usually ascribed to Seeley's famous phrase.[82]

In a similar way as Freeman's history for children, Seeley was able in *Expansion of England* to present his scientific views of history to a larger audience within the framework of a narrative of English history. While Macmillan was able to expunge much of the specific historiographical statements, in particular removing negative comments specifically directed against Green's *Short History*, a work that misunderstood the development of the British Empire in Seeley's mind, Seeley's general methodological and historiographical messages remained.[83] The book does not end by discussing the glory of the empire in English history, as one might expect from the popular presentation of the book as a mere product of the new imperial milieu, but rather with a rant against the popular notion that historians should make history more interesting. By the term 'interesting', Seeley argued, most people 'mean romantic, poetical, surprising'. He explained that he does 'not try to make history interesting in this sense, because I have found that it cannot be done without mixing it with falsehood'. But there is another meaning to the word interesting and that has to do with what 'affects our lives' and in this sense English history is inherently interesting. 'Make history interesting indeed! I cannot make history more interesting than it is except by falsifying it.'[84] Seeley was clearly able to put forward his scientific views despite Macmillan's heavy editorial hand.

Much like Seeley, the *English Historical Review* was also forced to flirt with attracting the common reader almost as a favour to its reluctant editor who was never wholly convinced by the journal's specialized nature. As we have seen (Chapter 5), the first issue of the *EHR* was published to much fanfare, though the *Saturday Review* seemed to understand the key issue in publishing scientific history when it reviewed the journal's first issue. After praising the founding of the *EHR*, the *Saturday* argued that '[t]he only doubt that can be felt as to the prosperity of such a periodical is the doubt whether it can obtain a sufficient number of regular subscribers, the public of book-buying students being unquestionably smaller in England than in any other of the great civilized countries of the world.'[85] The *EHR* was published under the assumption that there was a large audience not presently served by the more general periodicals, but it was also understood that the *EHR* would have to help create that audience. Quite early in the history of the journal it became clear that the audience was much smaller than originally assumed and that not enough new readers were being created to justify the journal's rather narrow publishing scheme. 'We are rather disheartened with the scanty sale of the Historical Review', Bryce wrote to Acton as the end of the first year of publication approached. Interest in the journal as well as subscriptions 'rather declines than grows'.[86] After the first few issues Longman wanted to see changes that would increase the readership.

During the first year of publication, Longman speculated that the journal was losing at least £30 an issue. Creighton believed that changing the entire 'character of the Review would not help' as the journal would certainly not be able to compete with the '*general* Reviews and Magazines; we can only exist in our special lines'.[87] Bryce, on the other hand, wrote to Acton that the journal 'seems needlessly heavy' but that Creighton believed this necessary for the journal to be 'scientific and special'.[88] If a general change in the character of the review could not be achieved, Longman suggested that perhaps the Royal Historical Society could help,[89] though Creighton certainly did not want the journal to have anything to do with the amateurish society that he believed was run by 'humbugs' and 'old fogies'.[90] It was decided that at the very least the payment of contributors would have to cease, at least temporarily, and that the journal could be well-served with the publication of the odd popular article.[91] This apparently was enough to convince Longman to continue publishing the money-losing venture.[92]

Creighton spent a considerable effort attempting to persuade W. E. Gladstone to write something for the journal's pages. He hoped that 'the publication of the third part of the Greville Memoirs might' spur Gladstone to write a review and perhaps 'bring before your mind many points, which have now become matters of history, on which your personal knowledge could throw much valuable light'.[93] He admitted to Gladstone that he believed that the *EHR* had done 'a great deal for improving English historical literature' but it was only too clear that there was not 'a sufficiently large public to take an interest in purely historical questions; and I am not surprised to find that there is not a large enough body to make the Review remunerative, but enough to enable it to pay its actual expenses'. Being unable to pay for contributors meant that the *EHR* was now at a considerable disadvantage when it came to making the journal more interesting. As he honestly explained to Gladstone, 'The combination of readableness and research is so difficult as to be almost impossible, and the ordinary monthly magazine can get the most readable historical articles by paying for them.' He hoped that a piece by Gladstone on a recent historical period would help make the 'Review of interest to the general public'.[94] When Gladstone agreed, Creighton wrote immediately to Longman to share the good news. 'May I tell you privately that I have extorted a sort of promise from no less a person than W. E. G. to writing something about the last instalment of the Greville Memoirs?' Creighton wrote to Longman. 'Surely after this you will not despair of the Review.'[95] Unfortunately, Gladstone's article did not improve the sales.

In a report probably written to Longman in 1888, Creighton wrote that the journal had proved it was 'not merely the organ of specialists' and that it 'has not neglected the consideration of the general public'. 'Each number', Creighton explained, 'has contained much that is of general interest'. Creighton certainly believed that the journal was still able to maintain a 'high standard of excellence' but the fact remained that the articles of general interest were not enabling the journal to meet its bottom line.⁹⁶ By 1889 Longman was quite ready to end his relationship with the journal and he forced Creighton to make a decision: either the journal would be published for the general reader and seek popular studies to fill the journal's pages; or else it would have to accept its role as an entirely specialized journal that would fill a very small niche market and hope that enough students and professionals of history would be willing to print their work in the journal for free. Creighton had no choice but to accept the latter framework or else undermine the perceived purpose and function of the review. The *EHR* would finally make it an official policy not to pay contributors and simply accept the fact that the best written works of history would appear elsewhere while the best researched but unattractive works would appear in the *EHR*.⁹⁷

At the very least the *EHR* provided a space for historians where they could concentrate on speaking with one another under a much more exclusive discourse and not worry about needlessly popularizing their narrative for the benefit of the general reader. This was, in theory, why the journal was created. But, the journal's closest supporters had visions of creating a larger readership, of using the review to mine a great untapped resource, and help establish that great hive of historical workers envisioned by Stubbs, workers who would in theory read and publish in the *EHR*. By the 1890s, however, this dream seems to have vanished. And one gets the sense that there was a general lamentation that the scientific historian could not make his work appeal beyond a handful of peers. Good research and 'readableness' was an 'almost impossible' combination according to Creighton, a comment that suggests that historians were beginning to mourn the days when they, in Acton's great phrase, used to 'project their own broad shadow upon their pages'.

7 THE DEATH OF THE HISTORIAN

And so you must have yet another 'eloge' in the E.H.R. ... I think the loss irreparable.

F. W. Maitland to R. Lane Poole, 30 June 1902

Edward Augustus Freeman died at Alcante in Spain on 18 March 1892. He had been seriously ill for forty-eight hours and, as described by John Evans to the Society of Antiquaries, he was struck down 'by a disease that in these days need not exist': smallpox.[1] Less than three years later, on 13 January 1895, John Robert Seeley would perish of a painful cancer that he had been suffering from for some time. In 1898 William Stubbs's health also began to fail. He preached his last sermon on February 1901 at Windsor in memory of Queen Victoria who had died the previous January, the same month that witnessed Mandell Creighton's death. On 22 April 1901 Stubbs took his final breath. That same year Lord Acton suffered a paralytic stroke. He withdrew from public life and eventually died on 19 June 1902. Within the space of ten years, the most outspoken proponents of scientific history were all dead.

When the *English Historical Review* was founded there was not much thought given to the writing of obituaries.[2] However, they would become a fairly regular occurrence in the journal's pages given the fairly high casualty rate suffered by the discipline in the final decade of the Victorian period. The first generation that so loudly sought to make history a professional discipline of study and implement scientific methods of research and writing had suddenly fallen silent. The journal that was born to serve that generation's vision of scientific history-writing quite abruptly found itself with the task of memorializing the discipline's self-appointed founders. Their obituaries significantly appeared in the article section of the journal and they were lengthy memorials of the historian's life-work and tell us much about their early legacies in the face of a moment where scientific history-writing had clearly peaked and was in decline. By also considering the obituaries that appeared in the periodical press, it is clear that this generation would be remembered for establishing history on a scientific footing while also maintaining the central and perhaps eternal artistic sensibilities of history-writing despite their claims to the contrary. Whereas this generation of historians sought to establish a significant break between their scientific endeavour and the more artistic one found in the writings of *belles lettres*, the writers of the obituaries often perceived more continuity than discontinuity between these approaches – and often lamented when such continuity could not be found. We can see through these obituaries not the sharp edges of difference, but rather the blurred lines of similarity between 'scientific' and 'artistic' history.

James Bryce was asked to pen Freeman's obituary for the *English Historical Review*, as he was considered one of Freeman's closest friends for the better part of thirty years. He was not uncritical of his friend, however, and much of the review can be read as a lament for the life Freeman could have lived and for the histories Freeman could have written if not for Freeman's odd eccentricities that would cause him to lash out for seemingly little reason or for his ability to simply ignore important subjects that did not fall directly under the specific line of his research. For Bryce, Freeman's was a life that could have been rather than a life that was.

Freeman's life was, argued Bryce, 'comparatively uneventful as that of learned men in our time usually is'. But it was not the events in Freeman's life that made an obituary necessary, but rather because 'the record of his life is … a record of his historical work and the journeys he undertook with it'.[3] Freeman's life may not be intrinsically interesting, but for historians, particularly professional historians, it was both interesting and instructive. Freeman was one of the first writers of history to devote himself entirely to that profession. Indeed, being a professional historian was Freeman's life-work.

However, as Bryce pointed out, because Freeman was unable to secure one of the coveted historical chairs until quite late in his life, he had to earn his living by publishing in the periodical press and it was in those writings, particularly for the *Saturday Review*, where Freeman's views concerning historical methodology would become known. Between the years 1860 and 1869, Freeman wrote 723 articles for the *Saturday*, and most were review articles.[4] It was through Freeman's relentless attacks on poor historical scholarship that many students of history learned what should be expected of the scientific historian.

One could not read a review written by Freeman without knowing precisely what he expected out of historical scholarship: accuracy and clear writing. 'It was this passion for accuracy and for that lucidity of statement which is the necessary adjunct of real accuracy that made him deal so sternly with confused thinkers and careless writers', wrote Bryce. Carelessness to Freeman was a 'moral fault', Bryce argued, because 'true conscientiousness' would necessarily lead to accuracy, whereas the inaccurate would no doubt be the result of an inherent weakness. This helps explain why he was so violently critical of writers such as Kingsley and Froude. 'Mere ignorance he could pardon, but when it was, as so often happens, even among persons of considerable literary pretensions, joined to presumption, his wrath was the hotter because he deemed it wholly righteous.' As Bryce explained, Freeman 'felt it his duty as a critic to expose imposters', and in this case Freeman provided a wonderful service 'to English scholarship second only to those which were embodied in his own treatises'.[5]

Freeman as a harsh critic of inaccuracy was an element of his identity that clearly stuck. Because of his outspoken criticisms many found him to be seemingly 'rough and prickly'.[6] But few doubted his accuracy. The *Speaker* argued that his work was distinguished by two merits: 'thoroughness' and 'accuracy'. 'He never slurred over a difficulty, he never spared any time or trouble that might need elucidate the problem.'[7] The *Athenaeum* concluded that Freeman's 'great attainments made him an *érudit* of whom England might well be proud'.[8] 'He brought a new element of minute accuracy into the region of history', argued the *Saturday Review*, 'banishing theories and conjectures, and placing the study to which he was most warmly devoted on a new and stable pedestal'.[9]

This obsession with accuracy was also necessarily an element of Freeman's writing style. The *Literary Opinion* explained that 'Mr Freeman showed that if we are determined to speak the truth about history, and to speak it intelligibly, we must busy ourselves with words and style'.[10] He expected, as Bryce would similarly explain, an exactness when it came to the use of words which to him 'appeared so essential to good work, that he was apt to set down the want of them rather to indolence than to incapacity, and to apply to them a proportionately severe censure'.[11] The *Literary Opinion* believed that Freeman chose his words not merely for clarity but that 'he desired beauty as well' arguing that his 'rules of style' were both sensible and reliable.[12]

Most observers, however, were not convinced that Freeman's obsession with accuracy had such a positive effect on his writing. Bryce had to admit that when it came to the issue of style 'the results were not wholly fortunate'. Freeman's obsession with exactness and thoroughness, Bryce believed, caused him to repeat 'the same idea ... with slight differences of phrase, in several successive sentences or paragraphs'.[13] The *Athenaeum* similarly complained that his books were plagued by an 'accumulation of detail which led to their being perilously long', while arguing that 'his style was injured by the fact that most of what he wrote was dictated as he paced up and down the room, and took little trouble in the way of subsequent correction'.[14] Most seemed to agree, however, that Freeman's rather 'diffuse' and 'repetitive' style was the result not of any inherent inability to write properly but of his method and love of history. 'The fault sometimes charged upon his writing', explained the *Speaker*, 'arose in large measure from the very warmth of the interest which he felt in his subject, and which led him to dilate, sometimes too fully, upon points that had no great attraction for the ordinary reader'.[15] Bryce believed 'that the prolixity we sometimes regret was due partly to his anxiety to be scrupulously accurate'. It was precisely because of the 'purity of his English', that his books were somewhat dry and unpopular, though Bryce was forced to admit that 'some of his friends', no doubt Bryce among them, 'thought he sacrificed too much to it'.[16]

If Freeman's obsession with clear language was deemed peculiar, equally so was the almost obsessive interest he would cultivate about the subjects he cared about 'and the scarcely less conspicuous indifference to matters which lay outside the well-defined boundary line of his sympathies'. As Bryce explained, 'If any branch of inquiry seemed to him directly connected with history, he threw himself heartily into it, and drew from it all it could be made to yield for his purpose.' But when it came to other subjects that lay just outside of his direct interest 'he cared so little that he would neither read not talk about them, no matter how completely they might for the time be filling the minds of others'.[17]

Despite the peculiar and specialized nature of his interests, he did expand the terrain of what could be considered proper historical research and evidence, particularly in the realms of architecture and geology. Bryce argued that Freeman believed that architecture was 'the prime external record and interpreter of history' and he spent much time analysing medieval architecture as an important source for his historical work on the subject. The *Archaeological Journal* particularly valued the 'technical knowledge' that he brought to the study of archaeology as well as the archaeological work that he brought to bear on his historical writings. The journal was particularly pleased at the way in which Freeman would visit and examine 'the scenes of such events as he proposed to describe'. The author also praised Freeman's deep knowledge of architectural and archaeological history: 'No man was better versed in the distinctive styles of Christian architecture, or had a better general knowledge of the earthworks from the study of which he might hope to correct or corroborate any written records, and by the aid of which he often infused life into otherwise obscure narrations.' As far as the *Archaeological Journal* was concerned 'Mr. Freeman was not only a great historian, he was also a great archaeologist'.[18]

In a similar vein, Freeman was also interested in the relationship between geography and history and because of that he was given a hagiographic memorial by the Royal Geographic Society. 'It is to Freeman, more than to any other writer of modern times, that the recognition is due of the necessity for a geographical training and geographical instincts in a true historian.' The Geographical Society greatly appreciated the light that Freeman's work shone on the discipline and the relevance of geography to both history and politics. 'He held that the physical geography of a country always has a great effect upon its political geography', C. Markham addressed the society, 'and that the historian has to deal with the nature of the land and with the people who occupy it, so far as political divisions have been influenced by them'.[19] Markham believed that Freeman 'has done most valuable service' to both geography and history by giving 'geography its true place in relation to the study of history' while promoting a scientific method that is central to both disciplines of knowledge.[20]

Notwithstanding the very warm memorials Freeman received from other disciplines, Bryce believed that Freeman did not appreciate the inherent value of those subject matters. When it came to architecture, he only appreciated it as it related to specific historical questions, but he did not appreciate architecture as art. He also had little interest in literature and when he did manage to pick up a book it was merely for the 'historical information' it might contain.[21] This was despite the fact that Freeman had a great interest in language. As was made most clear in his *Norman Conquest* (see Chapter 2), he enjoyed seeking the origins of words to help establish racial continuities, but much like architecture, philology was merely 'an instrument in the historian's hand'. Freeman 'took little pleasure in languages simply as languages – that is to say, he did not care to master grammar and idioms of a tongue, nor did he possess any aptitude for doing so'.[22] Freeman had 'a very peculiar mind', Bryce was forced to admit, 'which ran in a deep channel of its own, and could not easily, if the metaphor be permissible, be drawn off to irrigate adjoining fields'.[23]

This is not to say that Bryce was not forthcoming in his praise of Freeman. It is telling, however, that his positive comments were often backhanded. For instance, Bryce argued that Freeman 'loved to dwell in the past, and seemed to see and feel and make himself a part of the events he described', but this was despite the fact that he did not seem to have much imagination.[24] Indeed, even with Freeman's limited imagination, he had a great 'power of realizing the politics of ancient or medieval times'. For all his faults, Freeman did make history real.[25]

In spite of Freeman's obsession with accuracy, or perhaps because of it, and despite his prickly outward persona, a deep love of history underpinned all that he did. Unfortunately, this was a side of Freeman that one would not see by reading his books. 'He did not save himself for his books', argued the *Literary Opinion*. Instead, Freeman expended his great story-telling capabilities on conversations with his friends. His works may not have been terribly interesting, but he no doubt had 'a warm heart and a nature truly noble'.[26] Reference to Freeman's kindness and friendship, of his very warm heart, of his humour, was a general theme running throughout most obituaries. For the *Saturday*, 'Freeman was unsurpassed as a friend, and it speaks well for his character that it was those who knew him most intimately who loved him best'. The *Guardian* argued that the kindly nature of this supposedly gruff and angry man 'became ever more and more apparent' with age.[27] For the *Speaker*, he 'was essentially a warm-hearted and kind-hearted man, singularly loyal and constant to his friends, always ready to serve and defend them'.[28] It was simply unfortunate, Bryce would make clear, that he 'did not more frequently enliven his pages by indulging in the humour so natural to him'. Indeed, Bryce lamented that the general reading public did not get to see that side of Freeman that only his friends and colleagues saw. 'His letters sparkled with wit and fun, but it is only in the notes to his histories, and seldom even there that he allowed one of the most characteristic features of his mind to appear.'[29] Freeman was simply too good at suppressing his individuality and his personality when it came to writing history.

Seeley and Freeman were often mentioned in the same breath when discussing the state of history in Britain at the time when both were at the height of their powers in part because they brought forward a similar message about the importance of language and clear writing uncorrupted by literary style or subjectivity. They also shared the same dictum about the true meaning of history, that it was inherently about politics. 'History is past politics, and politics is present history' was a phrase coined by Freeman though Seeley certainly agreed with the dictum, so much so that a present-day historian who should know better recently and smugly criticized a postmodern feminist for supposedly wrongly attributing the phrase to Freeman, though of course she was right.[30] This almost blinding focus on politics as history was already viewed as narrow as early as 1886, as Bryce's 'Prefatory Note' for the *EHR* made quite clear.[31] But Seeley brought to the study of political history the methods of the scientist and if he could be criticized for the narrowness of his approach, he was heralded for bringing the proper tools to the study of his political histories.

J. R. Tanner (1860–1931) wrote Seeley's obituary in the *EHR* very much from the perspective of a former student celebrating the life and work of his favourite professor. Tanner was a former student of Seeley's and had come through the ranks at Cambridge. By the time of Seeley's death he had become one of the main historians of the English constitution, not quite as well regarded as Stubbs or F. W. Maitland but deemed an authority on the history of the subject nonetheless. Tanner focused primarily on Seeley's lectures and on his work as a professor of history students. Tanner's main purpose was to show that the same skill and hard work he brought to his publications, was also apparent in his lectures.

One of the other criticisms levelled at Seeley was that he engaged in 'hasty generalizations' in his lectures. He may have excelled in 'lucidity and thoroughness' but it was often said that he encouraged his students to discuss 'large considerations' that may have led them away from a specific focus on the facts. Tanner argued that nothing could be further from the truth. First of all, his lectures were a product of the same 'laborious process by which his finished work was produced'. Moreover, Seeley would not stand for imprecise language; he insisted 'first of all upon clearness of thought and expression'. He 'never permitted [his students] to wrap up fallacies in fine phrases, or to use high-sounding terms that had not been defined'. Indeed, just like Freeman, '[t]here was nothing that the professor enjoyed more than exposing this kind of imposter'. In his written work as well as in his lectures, it was very clear that Seeley 'hated above all things the picturesque in history'.[32] This theme in Seeley's work was noted by others as well.

Running throughout Seeley's other death notices, much like Freeman's, was Seeley's seeming hatred of style, a powerful villain in the scientific historian's discourse. For the *Academy*, Seeley was a member of that 'modern school' of history, 'which tends to sacrifice literary presentation to accuracy of research'.[33] In the words of the *Saturday Review*, Seeley was very much an 'accurate and painstaking historian, equipped with a scientific method, and a steady devotion to truth'. In this sense, 'Seeley's theory of history was the exact opposite of that of Professor Froude'.[34] Or, as Maurice Todhunter put it, the 'picturesque writing of Froude did not appeal to him much'.[35]

Even Seeley's lecturing followed a similar process of sacrificing style for truth. Tanner explained that Seeley kept a 'strong restraint upon himself, to concentrate deliberately his whole attention upon clearness, and clearness only'. Indeed, Seeley's lectures were given as if by a lawyer, consisting 'largely of dry statements of fact, marshalled ... with such skill'.[36] But when it came to his writing, many felt that that perhaps Seeley's 'disregard of picturesqueness or dramatic effect' was 'excessive'.[37] 'His fear that any sacrifice of truth should be made to picturesqueness or dramatic effect', argued the *Saturday*, 'undoubtedly made his books much less readable and interesting than they might easily have been'.[38] Much of his published work was simply unattractive and bordering on the unreadable.

Seeley's most impressive feat as a scientific historian was no doubt his *Life and Times of Stein* (1878), a laborious three-volume study that was, in the words of H. A. L. Fisher, 'so accurate, and well-informed, and discriminating, as to not only satisfy but to excite the admiration of the most exacting historical tribunal in Europe'.[39] But Seeley employed a writing style in the *Life of Stein* that could only be referred to as 'dry and unimpassioned' according to Tanner. And, as Tanner also made clear, 'the method of "Stein" was Seeley's ordinary method', one that certainly impressed other scientific historians but not so much general readers.[40] It was unsurprisingly his most unpopular book, and perhaps by design. For the *Saturday*, Seeley's *Life of Stein* was the prime example of Seeley's excessive hatred of the picturesque applied to an actual study of history where 'he seems to have deliberately resolved not to be popular'.[41] Todhunter believed that very few managed to get through the three thick 'unattractive' tomes though he heard that 'German judges have been known to praise it on account of its thoroughness and care'.[42] He also brought up Seeley's other unpopular work, *Natural Religion*. Many looked forward to the long-promised sequel to *Ecce Homo* and Todhunter admitted that he 'heard admirers of *Ecce Homo* speak disparagingly of *Natural Religion*', though as we have seen (Chapter 6), it was not meant to appeal 'to ordinary minds'.[43]

Despite the fact that Seeley seemed to, at times, employ a deliberately unattractive style, he published other books that, in the words of the *Academy*, 'prove that he possessed the saving grace of imagination. *Ecce Homo* and *The Expansion of England*, indeed, are, in their different ways, two of the remarkable productions of the later Victorian epoch.'[44] Fisher also admitted that despite Seeley's desire to remain unpopular 'there is something extraordinary in his literary career'. He also pointed to these two very different books as leaving a profound mark on Victorian society. 'Twice he took the English reading world by storm, once by a book on religion, and again by a book on politics; and each book, in its own sphere, may be held to mark an epoch in the popular education of the Anglo-Saxon race.'[45]

A further irony in the popularity of *Ecce Homo* is that Seeley never claimed ownership of the book. The *Academy* argued that he was appointed Regius Professor because 'he was chiefly known as the author of *Ecce Homo*, though we believe that he never acknowledged the paternity'.[46] The *Saturday* pointed this out as well arguing that *Ecce Homo* 'was never, we believe, deliberately acknowledged by its author, though the authorship was soon an open secret', this despite the fact that it was *Ecce Homo* that really 'stirred the minds of [Seeley's] contemporaries'. For the *Saturday*, *Ecce Homo* was brilliantly written and proof 'beyond question' that the 'absence of brilliancy, of picturesque and telling effects' in Seeley's other works of history 'was the result of deliberate repression, not of any lack of literary faculty.'[47] Indeed, Seeley proved that he could write with style and excite the masses if he chose to do so.

Tanner decided to let the general presses rehash the 'half-forgotten controversy that raged round "Ecce Homo" and the well-kept secret [!] of its authorship', and instead focussed on the works Seeley actually admitted writing. Like the other obituary writers, Tanner also pointed to *The Expansion of England* as a profound success and he argued that it 'in spite of [his] resolute self-restraint, Seeley ... was keenly alive to the poetry of history, and when he chose, the effect was irresistible.'[48] It was not just that *The Expansion of England* 'was eminently seasonable' when it was published, though that certainly had something to do with the book's popularity according to the *Saturday*.[49] It also clearly 'appealed to the pride and patriotism of most English-speaking persons'. But it also, says Todhunter, discussed the 'great question' of the British Empire 'in a new and original manner.'[50] Tanner pointed out that it must be remembered that Seeley's story of the British Empire 'was first told in a Cambridge lecture-room, and told in such a way as to stir the imagination and quicken the pulses of the dullest undergraduate among the audience'. It was essentially a romantic story of how the English nation and the British Empire were all a part of the same whole, that you could not truly have one without the other. For Tanner, 'Seeley's conception of empire was the conception of a poet as well as an historian'.[51]

Tanner was not unaware of his old professor's hatred of the poetic in history as it was so clearly tied to his disgust with popular and Romantic history in general. '[I]n spite of his apparent renunciation of rhetoric', Tanner argued, '[Seeley] was keenly alive to the rhetorical possibilities of his subject'.[52] It should be clear that Tanner's appreciation of Seeley was at odds with Seeley's own notion of what made for solid historical scholarship. Tanner even believed that it 'matters little to his memory' whether 'Seeley is right or wrong in his view of history'. What was most important, Tanner argued, was to realize that Seeley 'was a great influence in his day and generation in favour of thoroughness of investigation, of habits of clear thinking and lucid expression' and, most importantly, 'he did all in his power to bestow upon his pupils the incommunicable gift of style'.[53]

Much like Bryce's obituary of Freeman, Tanner referred to aspects of Seeley's historical work that might not have been noticed by the casual reader and, indeed, might not have been noticed by Seeley himself. Whereas Bryce lamented that Freeman's personality was not better portrayed in his work, Tanner suggested that it was Seeley's personality, his style, that made his work so remarkable. This tends to contradict one of Seeley and Freeman's most important methodological principles, that it was precisely the historian's subjective presence, his personality, his style, that had to be eschewed in the writing of professional history. By the time of Stubbs's death at the turn of the century, however, it is apparent that the views of Tanner and Bryce were not part of a heterodox movement from within the discipline but, rather, an orthodox view that was shared by members of a new generation of historians. This would become even more clear six years later when the reminiscences of Stubbs's life would begin to appear.

When F. W. Maitland (1850–1906) was approached by R. Lane Poole (1857–1939), then editor of the *EHR*, to write Stubbs's obituary for the journal, Maitland's immediate response was no. 'I cannot (I am glad to say it) plead ill health', Maitland wrote to Poole. Nor would Maitland 'plead other engagements', effectively denying himself the majority of excuses editors are forced to accept. However, Maitland did want to write the obituary. Indeed, he was flattered to have been asked: 'I feel that I owe so much to the good bishop and I so deeply admire him that I should not like to put in any such pleas as the above if the editor of the E.H.R., conscious of his responsibility, asked me to write.' And yet Maitland was deeply distressed at the prospect of putting some words to paper about Stubbs.[54]

As we have seen with Freeman's and Seeley's obituaries, their personalities were the focus as much as their historical work. Their identity outside of the name that was attached to their historical studies was, in a sense, brought to life. It was precisely this aspect of Stubbs that Maitland could not create.

Stubbs was to me simply and solely that writer of certain books. I never spoke to him. I never saw him but once, and that in church. Therefore I could give none of those 'reminiscences' which people expect, and naturally expect, on these occasions. Not a word could I say of the man as distinguished from the writer – no word of his kindness and geniality or the like – and yet I should suppose from vague report that many words of this sort could be written and are in some sort due to him. ... Do you think that it would seem strange to friends and pupils if you put this article into the hands of one who will speak of Stubbs merely as he would speak of Hallam or of some other writer of books? I should (so I think) speak warmly – but obviously what I said would come from the merest stranger, and indeed I should be compelled to say in the openest manner that the man and the bishop I knew naught.

It is not that Maitland did not want to 'appraise the merits of this very big man' but rather that he felt he simply did not have the personal authority to do so. Maitland was concerned that there would be people who would 'say (and not without truth) that it is like my impudence'. Maitland basically gave in to the request when he suggested that Poole could turn to him as a last resort; that is if, 'for one reason or another', Poole could not 'get an article from a friend'.[55] This was precisely the problem, however.

Stubbs had been out of active historical duty after he accepted a bishopric and gave up his chair at Oxford, in 1884, two years before the *EHR* was finally in production. Anyone training as a 'historical worker', as Stubbs would have put it, after 1885 likely would not have even met Stubbs. Moreover, the closest friends of Stubbs were dead or dying. His closest friends, J. R. Green and Freeman, had already gone. Lord Acton, quite likely Stubbs's biggest admirer and closest intellectual rival, was literally on his deathbed. Maitland made this clear to Poole in the last line of his letter concerning Stubbs's obituary: 'I very gravely fear that our woes are not at an end. I hear very bad news of Acton.'[56] Stubbs and Acton were the last of a generation of scientific historians. Poole, therefore, had to rely on a historian who was at least familiar with Stubbs's work, and the choice of Maitland was perfect in this regard.

Maitland was the biggest name of this second generation of professional historians. Much of this had to do with the fact that his primary work, *The History of English Law* (1895), was partly a history of the English constitution, the same subject as Stubbs's well-known *Constitutional History of England* (1873–8). When Stubbs's work appeared it immediately became the most important work of English history, the symbol of disinterested historical observation based on extensive archival research, cementing him as – by any professional historian's account – the greatest English historian. Maitland's work was also well received, praised by professionals and not least of all Stubbs, and Maitland was considered the new torch-bearer of the high historical standards set by Stubbs and his generation. This was no better symbolized than in the fact that Maitland's history of English law was on the reforms of Henry II and the Magna Carta of 1215, an obvious extension of Stubbs's history of the constitution which centred on the Norman Conquest of 1066. And Maitland's history of the constitution was more authoritative and more 'real' than Stubbs's who had already effectively stabilized the meaning of the unwritten constitution.[57] Perhaps it is not surprising, then, that Maitland wrote Stubbs's obituary on behalf of his own generation, a generation who only knew Stubbs as a name, the biggest of names, attached to history books.

'No readers of the "English Historical Review"', Maitland began, 'no English students of history, no students of English history can have heard with indifference the news that Dr. Stubbs was dead. A bright star had fallen from the sky.' Instead of avoiding discussing the personality of a bright star that he did not really know, Maitland made clear his relationship with Stubbs and the limitations of this obituary:

> This is not an attempt to speak on behalf of those who, without being his close friends, yet knew him well. Evidently there is much to be told which only they are privileged to tell of a man who was good as well as great, of a kindly and generous, large-minded, warm-hearted man. Then there is the bishop to be remembered, and the professor, the colleague in the university, and the counsellor of other historians, whose ready help is acknowledged in many prefaces. Evidently also there is something to be added of good talk, shrewd sayings, and a pleasant wit. Of all this some record has been borne elsewhere, and fuller record should be borne hereafter.

'But to this journal', Maitland argued, 'there seems to fall the office of endeavouring to speak the grief of a large but unprivileged class'. And here Maitland spoke on behalf of his generation, on behalf 'of those to whom Dr. Stubbs was merely the author of certain books, but who none the less cordially admired his work'. The death of Stubbs was a huge blow for this unprivileged class because, as Maitland made clear, in Stubbs 'they have had a king and now are kingless'.[58]

The *Athenaeum* spoke in similarly apocalyptic overtones, arguing that the 'death of the Bishop of Oxford has deprived the learned world of its greatest medieval historian'. There was no doubt in this author's mind that 'the late Bishop's record has no complete parallel in any other country' because he had 'traversed the whole ground of a nation's medieval history in the spirit of scientific research'.[59] Maitland was clearly not alone in believing that Stubbs was the king of history.

The work that was often invoked as symbolic of the scientific framework Stubbs brought to the study of history was, of course, his 'difficult but valuable' *Constitutional History*.[60] It became the example of a proper inductive method of historical science at work, a 'great work' that, according to the *Athenaeum*, all serious students of history must have 'made a more or less intimate acquaintance with'.[61] It was such 'a mass of knowledge at first hand', reminisced a writer in the *Spectator*, 'such a marshalling of facts, such weighed and weighty sentences, such a judicial mind!'⁶² Maitland sought to personalize his relationship with Stubbs by explaining his own intimate relationship with the *Constitutional History*. Maitland explained that he believed he had an 'advantage over most of its readers'. This is because he was not forced to read it as so many students were before him. He actually came across the book 'in a London club, and read it because it was interesting'. It was so interesting, in fact, that Maitland did not read it critically but rather fell 'completely under its domination' accepting completely Stubbs's interpretation while only much later finding minor bones of contention.⁶³

Maitland's own historical identity was very much tied up with Stubbs's *Constitutional History* and this was no doubt true of Maitland's generation. They were taught that this was the best work of history ever produced by an English historian.⁶⁴ Tanner claimed that 'to read the first volume of Stubbs was necessary to salvation; to read the second was greatly to be desired'. The third volume, however, 'was reserved for the ambitious student who sought to accumulate merit by unnatural austerities'.⁶⁵ Indeed for most historians Stubbs's work was vastly important and therefore necessary to read and to study, but it was a difficult journey to get through the entire work.⁶⁶ Yet Maitland came across the work almost by accident and read it with pure enjoyment, falling 'completely under its domination'. He appreciated the book for something Stubbs and his scientifically oriented colleagues had tried to expel from history-writing: its art.

For Stubbs and his contemporaries, art was the antithesis of professional history. Art was that which corrupted the work of historians such as Macaulay and Froude. They were more interested in telling an interesting story than in giving an accurate portrayal of what really happened. He could find nothing of value in such work. 'His scorn of popular history was complete', argued the *Academy*, and he along with Freeman 'pilloried Charles Kingsley and Anthony Froude' for their artistic historical lies.⁶⁷ It was certainly true that, according to the *Athenaeum*, Stubbs's scientific reputation was 'largely due to his appreciation of the grave defects of our own national system of historical study and to his readiness to avail himself of the results of the scientific methods of research employed by foreign scholars'.⁶⁸ Indeed, from the first words that Stubbs uttered as Regius Professor at Oxford, he made it clear that he wanted to found a school of history that was filled not with a handful of great individual men who would produce magnificent works of art but rather with a 'republic of workers' who would replicate the work done by the historical workers in Germany. Stubbs wanted to be looked upon 'as a helper and a trainer': 'I desire to use my office as a teacher of facts and the right habit of using them.'⁶⁹

For Stubbs, acquiring and presenting the facts of history was hard work, something that Romantic historians as well as readers of such work simply could not understand. The first sentence of his *Constitutional History* is telling in this regard: 'The History of Institutions cannot be mastered – can scarcely be approached – without an effort.'[70] This not only meant that his own methodical process in collecting and then disseminating historical knowledge was difficult but that reading such a work should also be a painful exercise. The *Academy* certainly concurred with just this point. 'As an historical anatomist Dr. Stubbs had few rivals', argued the *Academy*, 'but he rarely, if ever, could be enjoyed by the general reader', and this was likely by design rather than by accident.[71] Stubbs would have been dumfounded by Maitland's joy in what should have been a difficult and painful learning experience. The *Athenaeum* argued quite rightly that Stubbs's 'own definition of "historical genius" was "an unlimited capacity for taking pains," and there was certainly no limit to his own capacity'.[72] He was, apparently, a sadistic king.

Maitland certainly knew of the difficulty in producing the kind of history that Stubbs had written. He believed that there was nothing inherently interesting about the study of institutions, and he argued that Stubbs had a deep conviction in the necessity in writing about their history. The history of the constitution, for Stubbs, 'is the absolutely necessary background for all other history, and ... until this has been arranged little else can be profitably done'. This was an unfortunate but necessary task, argued Maitland, and Stubbs was always willing to engage in the difficult task of history when 'duty called'.[73]

Duty actually called much earlier, in 1857, when the Rolls Series was planned and Stubbs was hired on to edit and catalogue the series. It was well understood that editing the series provided Stubbs with professional credentials at a time when record officials at the Public Record Office rather than 'the ill-organized university historians' were actually sifting through the recorded traces of the past.[74] '[I]t should also be remembered', argued the *Athenaeum*, that Stubbs 'had first served an apprenticeship (so to speak) as an investigator of original sources in connexion with his memorable editions of medieval chroniclers in the Rolls Series'.[75] Maitland made it clear that editing the Rolls Series was a fairly thankless task that not many were even competent enough to take on, and those who were, were unlikely to be interested: 'The picture of an editor defending his proof sheets sentence by sentence before an official board of critics is not to our liking.' Thankfully, Maitland argued, Stubbs heeded the call to duty and 'raised the whole series by many degrees in the estimation of those who are entitled to judge its merits'.[76]

It was the introductions to the seventeen stout volumes that Stubbs had edited which Maitland most appreciated. 'He was ... a narrator of first-rate power: a man who could tell stories ... in sober, dignified, and unadorned but stirring and eloquent words'. Unfortunately, Maitland argued, the general reading public never got to appreciate Stubbs's 'artistic merits' as his edited volumes and *Constitutional History* were for historians' eyes only. But it was precisely the artistic merits of Stubbs's writing that Maitland spent the majority of the obituary recounting. Indeed, it was by examining Stubbs's written work that Maitland managed to draw out some elements of Stubbs's personality, a personality that Maitland could only know through Stubbs's writing.

'A word too should surely be said of the art' of Stubbs's historical work, Maitland explained, 'unconscious art' he had to admit, 'but still art'. It was art that maintained the reader's interest, Maitland argued, it was art that 'kept us reading'. For Maitland the *Constitutional History* had to be considered a work of art: 'It is so solid and so real, so sober and so wise; but also it is carefully and effectively contrived.'[77] The *Spectator* was also impressed with Stubbs's style of writing, though admitted to preferring Stubbs's introductions to the Rolls Series 'where, with more room, the master let himself go'. What was so impressive about the writing of *Constitutional History* for the *Spectator*, however, was Stubbs's 'power of restraint', a power that is only 'possessed by the greatest men'. As the author pointed out, '[n]ot until the last paragraph of the third volume of his greatest work comes any personal expression'.[78]

Maitland was also impressed with the way in which the book acted as an example – in the very way in which it was written – as to how the scientific historian was to do his duty. Stubbs's work was, according to Maitland, a 'practical demonstration' of the 'historian's art and science':

> No other Englishman has so completely displayed to the world the whole business of the historian from the winning of the raw material to the narrating and generalizing. We are taken behind the scenes and shown the ropes and pulleys; we are taken into the laboratory and shown the unanalysed stuff, the retorts and test tubes; or rather we are allowed to see the organic growth of history in an historian's mind and are encouraged to use the microscope.[79]

For Maitland, Stubbs was engaged in scientific experiments; he analysed the facts, the 'stuff' of history. However, it was only Stubbs's use of art that could have allowed the reader to become a virtual witness to his historical experiments, it could only be art that would allow the reader into 'an historian's mind' and be 'encouraged to use the microscope'.[80] Perhaps the *Athenaeum* had Maitland in mind when it argued that 'a later generation' of historians 'has discovered the art of reading' the *Constitutional History*.[81]

Needless to say, Stubbs would have taken it as a profound criticism to suggest that he had employed anything but science and hard work in his historical writing. The type of individual creativity that is invoked by the notion of art was precisely the opposite of what Stubbs believed history should be. History was to be a grand collective endeavour, with the whole of the discipline naturally being greater than the sum of its parts. And yet when Maitland thought about Stubbs and his generation of fellow scientific historians, he lamented the passing of the age when the historian was still bigger than the history, quite in contrast to the image Stubbs presented of himself at his inaugural lecture at Oxford.

Stubbs's death, it should be noted, was preceded only a few weeks earlier by Mandell Creighton's, which was also viewed as a significant blow to the discipline of history in Britain.[82] It is clear that Creighton's death as well as Acton's impending death, had Maitland musing about the recent casualties of the discipline, about the loss of an entire generation of historians. '[O]ur thoughts may naturally go back to the year 1859', Maitland argued, 'when Hallam's death was followed by Macaulay's'. And like Hallam and Macaulay, 'we may think of [Stubbs and Creighton] as belonging to a past and a remarkable time'.

> Was there ever, we might ask, any other time when an educated, but not studious Englishman, if asked by a foreigner to name the principal English historians, would have been so ready with five or six, or even more names? Freeman and Froude, Stubbs, Creighton, Green, and Seeley he would have rapidly named, and hardly would have stopped there, for some who yet among us had already won their spurs.

By linking the deaths of Stubbs and Creighton with that of Macaulay and Hallam, Maitland ignored that dividing line that supposedly separated the scientific from the Romantic, the professional from the amateur. For Stubbs – as well as for Creighton, Acton, Seeley and Freeman – the purpose of making history a science was not to create another group of great individual historians, but to create a body of historical workers who would be trained to suppress their subjectivity and present the past as it actually happened – to let the past speak for itself. Indeed, if anything, Stubbs's death should have been viewed as the logical fruition of the scientific historian's striving to overcome subjectivity. Only in death could the self truly be transcended.[83] Yet for Maitland, Stubbs was a great historian in the same vein as Macaulay; his death meant that the discipline of history was now without a king. Less than a year after Maitland's article on Stubbs appeared, the last great English historian, Lord Acton, would be dead. 'And so you must have yet another "eloge" in the E.H.R.', Maitland wrote to Poole. 'I think the loss irreparable.'[84]

Poole decided that he should write Acton's obituary for the *EHR*. He had been involved with the journal from the beginning, acting as sub-editor under Creighton until 1895 when he became joint editor with Samuel Gardiner. He would eventually become sole editor in 1901 and would hold that position until 1920. He and Acton became friends through their work for the *EHR* as their fairly lengthy correspondences held at the Acton archives can attest.[85] Acton clearly respected Poole's judgements about his own work. When he was trying to republish his very scattered historical articles and reviews in a single volume, he asked Poole for advice on the pieces to include.[86] It was also to Poole that Acton turned when he needed help determining the authors to approach for the *Cambridge Modern History*. After sending Poole a list of possible candidates Acton wanted to know: 'Whom do you strike out? Whom do you add?' He also wanted Poole to let him know 'what each man is good at'.[87] Poole's lengthy work for the *EHR* gave him great insight into the entire discipline of history and he was certainly well placed to judge Acton's place within the community.

For Poole, Acton's death was a truly momentous occasion, one he felt was not being appropriately memorialized in the periodical and newspaper press. 'That with Lord Acton England lost the last of a generation of great historians has not been fully recognized in the obituary notices which appeared after his death on 19 June [1902].' A generation of 'great historians' was now gone, argued Poole, and he felt it necessary to highlight that fact in his review of Acton's life and historical work. Poole recognized that for many Acton's work as a historian was only one element of a life that was embroiled in religious controversy and political work. Poole's obituary, however, would be 'solely concerned with his work as an historian'.[88]

It was not until very late in Acton's life that he devoted himself entirely to the study of history. 'He never', Poole believed, 'until he was sixty, delivered a lecture'. However, as Poole made clear, Acton threw himself into anything he was interested in. When he became interested in history, he was completely devoted to it. He did, however, edit the *Rambler* for several years, publishing pieces on a variety of topics. Criticizing works of history in the *Rambler* became Acton's specialty and he became known as someone with a vast knowledge of all the primary and secondary literature of many historical topics. 'It is this fullness of equipment that strikes one first of all in reading any criticism by him', Poole argued. 'While he never assumed the pose of superiority, but rather seemed bent on showing how respectfully he appreciated the merits of the author whom he treated, the impression left was that he wrote from a higher level of knowledge and that he had it in his power to support or demolish more than he thought appropriate to the present purpose.'[89]

The vastness of his knowledge was strongly noted in the other obituaries as well. For G. P. Gooch, writing for the *Speaker*, Acton had an 'unrivalled knowledge of books ... and men'. He quoted the famous saying of Henry Sidgwick that 'however much you knew about anything, Lord Acton was sure to know more'.[90] He was an avid reader and collector of books. As Poole explained, he 'read unceasingly and in the widest range, and he was reputed to be the best-read man of his time'.[91] At the time of his death, his library was estimated to contain 60,000 volumes.[92] Others noted the cosmopolitan nature of his education. 'He thought in German and in French as easily as in English', argued the *Athenaeum*, 'and spoke these languages as perfectly as any man in the world can speak these languages'.[93] The *Edinburgh Review* went so far as to compare him to Erasmus and their shared 'cosmopolitan culture, the thirst for knowledge'.[94]

For Poole, the most important aspect of Acton's education was his training at the feet of the German masters. He noted that Acton 'had attended the lectures of Boeckh, Ranke, Riehl and many more of the same generation' not least of which was Döllinger whom 'he knew on terms of the closest of friendship'.[95] Similarly, the *Edinburgh Review* argued that Acton clearly 'profited by his stay in Germany. He always described himself as a disciple of Ranke, and never let the opportunity pass of keeping up relations with him, with Droysen, Sybel, Bluntschli, Roscher, Hermann, Sickel, Dilthey, and others.'[96] Indeed, Acton's firsthand knowledge made him the unofficial English interpreter of the German school of history at the moment when professionalizing historians in England were looking to Germany as a model to follow.[97]

Much like Seeley, Freeman and Stubbs, Acton cultivated a persona whereby the 'accumulation of facts and the accurate presentment of affairs' was a moral duty for the historian. Indeed, as the *Athenaeum* made clear 'he set the highest store on impartiality' but as Gooch pointed out he also 'insisted on the duty of applying the moral standards to the actions of personalities of the past no less than of the present.'[98] As we have seen (Chapter 5) Acton's deeply held concept of objectivity was impressive but it was also a burden that weighed on his ability to speak on behalf of the discipline as a whole, for he knew his belief in a transcendent universal moral code was unique. He refused to write the inaugural editorial statement for the *EHR* and seems to have had similar qualms about writing the same for his own planned *Cambridge Modern History*. He often refused to put himself in a situation where his disinterested persona might be compromised. As Maitland made clear in the *Cambridge Review*, Acton had an 'acute, an almost overwhelming sense of the gravity, the sanctity of history'.[99] Perhaps because of this, as Gooch explained, 'when he took a pen in his hand the stream of inspiration ran dry'.[100]

Acton's lack of historical production was a focus of much speculation in every obituary and was particularly addressed by Poole. Poole was adamant that Acton clearly had the knowledge to complete any historical study he set his mind to and he certainly had the opportunities, arguing that 'when urged to write a book which should take its place in a famous German historical series, [Acton] resisted the pressure'. By pressure, Poole was partly referring to the pressure Acton received from the publisher. He was also, however, referring to the pressure of the nascent discipline, the pressure to publish a specialized and primary source-based monograph, like that of Freeman's *Norman Conquest* or Stubbs's *Constitutional History*. Acton managed to resist this pressure his entire life: 'He never wrote a book.'

> His writings must be sought out, if that is possible, in the two magazines which he edited as a young man, and in the compositions of his mature age which he published in various periodicals, and of which some of the most characteristic specimens

appeared in the earlier volumes of this Review. When to these we have added the introduction which he contributed to Mr. Burd's edition of the 'Principe' of Machiavelli, and his inaugural lecture at Cambridge, we have nearly completed the tale of the printed work of his name.[101]

Gooch also pointed out that the entirety of Acton's published writings 'could be compressed within a couple of volumes'.[102] The *Athenaeum* speculated that what kept Acton from writing was the fact that he knew too much. 'He knew far too much to write books, for on every topic a thousand authorities would suggest themselves, and consequently the drag of hundreds of problems.'[103] For the *Academy*, it was a combination of his vast knowledge with a seemingly impossible methodology that led to Acton's literary paralysis. 'Lord Acton was a great scholar', argued the *Academy*, 'whose very weight of learning and scholarly love of accuracy hampered his production'.[104] Despite that lack of production, however, Gooch believed that those few published fragments 'possess unique value, and reveal their author scarcely less clearly than larger works might have done. ... No historian has written so little history; but few men have inspired so many books.'[105]

Of those few published materials, the *Edinburgh Review* particularly highlighted Acton's lectures which 'are his masterpiece'. Those lectures, the *Edinburgh Review* continued, 'when compared with famous addresses delivered at universities and academies, may be considered as unsurpassed'.[106] Indeed, they not only disclose 'an encyclopaedic knowledge' but they also indicate 'a truly great art. The style is clear, vivid, eloquent, and of original distinction.'[107] The *Athenaeum*, however, was less impressed with Acton's writing and lecturing style. The writing suffers partly from a 'cosmopolitan flavour, partly, also, to the overcrowding of materials which congested the flow of his exposition'. His writing often 'so bristles with quotations and references that we are diverted from the argument to wonder at the learning of the writer'. His lectures suffered from the same problem. 'They were so packed full of matter as to be obscure and difficult to follow.'[108]

Poole completely disagreed with the *Athenaeum*'s assessment of Acton's writings and was more in line with Gooch's and that of the *Edinburgh Review*. Poole believed that Acton's writing was always clear and extremely concise and yet profoundly thought provoking. 'He always chose the right word, the happy phrase; and his simplest sentences told because they were charged with more meaning than first caught the eye.'[109] Poole was adamant that Acton had a wonderful style of writing. Upon reading the obituary, Maitland wrote to Poole with approval of his assessment of Acton's life and work. As we have seen, Maitland was also commissioned to write an obituary for Acton and was impressed that both he and Poole focused on Acton's methodology as none of the other notices of Acton's death made any mention of it. 'I wrote of it somewhat noisily', said Maitland, 'because there was an impression here [Cambridge] that A. was a sort of Dictionary of Dates'. Maitland was particularly pleased with what Poole said 'of A.'s style – it is just what I wanted to see said'.[110] They both would have likely agreed with the *Edinburgh*'s assessment, that Lord Acton 'was truly an artist: his work of art was his life'.[111]

In 1886 it was the scientific spirit that led historians to found the *English Historical Review*. In the lead 'Prefatory Note', the all-important editorial statement of the premier issue, it was made clear that only history written by a guiding science would be published. It is clear from the obituaries analysed here that it was believed the four historians did much to establish history itself under a scientific framework. They were all praised for their unflinching devotion to accuracy, to impartiality, and to their very careful use of language. They were also duly thanked for bringing the historian's inductive scientific method, the devotion to impartiality and disinterest, to a wider community of students and historians alike. The obituaries can be read as the coming to fruition of the scientific history that they all worked so hard to establish.

However, it is possible to read into these obituaries a somewhat different story. They tell a story of a generation fascinated with the great names of history, with their lives, with their labour and, most importantly, with their personalities. They tell the story of a generation that no longer viewed science and art as mutually exclusive concepts, finding in the founders' works, traces of artistic merit. This was most pervasive in Seeley's obituaries where Tanner in particular found Seeley's formulation of British imperialism poetic, despite Seeley's clear denunciations of the literary in historical narratives. This was less pronounced in the obituaries of the other three, but Poole, Gooch and the *Edinburgh Review* praised Acton's style, Maitland found Stubbs's writing filled with personality, and Bryce appreciated the sense of reality that filled Freeman's works. It was the art of the historian's science that this second generation seemed to cherish.

There is also a clear sense that the authors of the obituaries did not entirely appreciate the identity that was created out of the inductive and disinterested demands of the historian as scientist. Bryce as well as the *Saturday Review* were disappointed that Freeman's published work was so repetitive and specialized and that it did not 'sparkle' with the 'wit and fun' that characterized his personality. Poole and the *Academy* portrayed Acton as practically crippled by his own methodological demands. Maitland lamented the specialized audience targeted by the 'unconscious art' of Stubbs's storytelling abilities, while Tanner and the *Academy* felt that Seeley's hatred of the picturesque was simply excessive leading the man to write a few unreadable books. It is actually quite shocking to read these obituaries alongside the methodological pronouncements of the deceased. They were praised for that which they denounced in others and in some instances they were criticized for adhering too strongly to their normative scientific methodology.

By the time this generation of historians died, something had clearly changed in the historian's communal discourse. A blind adherence to the science of history was no longer the primary attribute associated with a professional historian's identity, an identity that could now be devoted to both the science and the art within the historian's practice. This shift is no better illustrated than in a 1907 dinner given in Poole's honour to celebrate the *EHR*'s twenty-first anniversary. Bryce, who was the elder statesman, presided over the dinner held in Balliol Hall. As we have seen, as young man Bryce became friends with both Freeman and Stubbs. Later in his life he also befriended Lord Acton. He also knew Seeley quite well. It was Bryce who penned the initial editorial statement twenty-one years earlier so there was a sense that the *EHR* had come full circle when Bryce began to speak. Bryce, however, had more than the last twenty-one years in mind. 'The question is often asked', he said, 'whether History is a science or an art. You might as well ask whether the sea is blue or green. It is sometimes the one and sometimes the other.'[112]

EPILOGUE: FROUDE'S REVENGE

> Thus historians are between Scylla and Charybodis, to use a novel phrase. They jump, like Mr. Froude, into a sea of MSS and bring up a book of absorbing interest – a pearl, but a bizarre pearl, like those cunningly set in gold by the artists of the Renaissance. Or they pour over their work with a patent double-million magnifying pair of spectacles, and never produce anything worth looking at. Of the two maladies give me Froude's disease. Measles is better than paralysis.
>
> Andrew Lang, 'History as She Ought to be Wrote' (1899)[1]

Even though there is clear evidence that art was once again becoming an instrument of the English historian's toolkit, and in the pages of the specialized *English Historical Review* no less, the profession continued its boundary work against Froude with varying degrees of success, though ending in ultimate failure. The same could certainly not be said for Henry Thomas Buckle whose presence disappeared from the discipline of history in England almost as soon as the man died of typhoid in 1862.[2] But the violence against Froude's work spanned his entire adult life, continuing into the 1890s and even beyond. His work and method would eventually be vindicated but only well after he passed away when the work of his great scientific contemporaries began to be judged from more critical eyes.[3]

While Froude was never welcome to publish in the pages of the *EHR*, his work was still derided there, most notably in a pair of reviews that appeared in 1892 and they read as if penned by Freeman's dead hand. Augustus Jessop began his review of Froude's *Divorce of Catherine of Aragon* (1891) by simply noting the unfortunate fact that Froude ever decided to become a historian in the first place. 'It was a calamity to himself', argued Jessop, and, more importantly, 'it was a great misfortune to English historic literature, when Mr. Froude, nearly forty years ago, became possessed by that historic delusion which he has never been able to shake off, of which he is now the unhappy victim, and which, like all fanatics, he is passionately desirous to impose upon all who will listen to his pleading'. His thirty years of research, Jessop explained, have simply been wasted upon him. Jessop, much like Freeman before him, argued that ideally Froude's work would merely be ignored, particularly by the historical profession. However, because of Froude's 'exquisite style', 'it is only right that Mr. Froude's paradox should be exposed again and again ... while the sophistry or the inaccuracy of so charming a writer should be pointed out to the unwary'.[4] He concluded his review by referring to Froude's mistakes as 'perversions'. He went on to say that it 'would be unfair to condemn them as dishonest, but they are the inevitable outcome of a method of writing history which is, to say the least indefensible'. That method of history is of course one based not on facts, but on the desire to entertain a popular readership through beautiful prose while perverting the true character of the past.[5]

Martin Hume's review of Froude's *Spanish Story of the Armada* (1892) was less violent but offered a similar appraisal of the historian whose work simply could not be trusted. He, of course, admitted that Froude 'painted a brilliant and vivid picture of the most dramatic incident in English history'. That story is 'interesting, striking, theatrical' but, sounding much like Freeman on Froude's 'Thomas Becket', Hume was adamant that 'it is not history nevertheless'.[6] Again, it is the 'author's method of writing history' that is deemed the problem. Froude simply sought to present 'very dramatic' pictures that are 'unwarranted by Mr. Froude's own authorities'.[7]

That Froude once again felt the wrath of the historical community in 1892 is significant because that is the year he succeeded Freeman as Regius Professor of Modern History at Oxford.[8] The appointment, much like that of Kingsley at Cambridge thirty years earlier, was highly controversial. Many historians were shocked that a seventy-four-year-old Froude would be appointed instead of well-respected practitioners of scientific history such as Samuel Gardiner and F. York Powell.[9] The appointment spoke to the fact that despite the growing cohesion, strength and consensus of the historical community, appointing of the most symbolic and powerful historical positions was still in the hands of politicians who could be swayed by a variety of factors. Then-Prime Minister Lord Salisbury, who happened to enjoy Froude's histories and was also his friend, ignored the recommendations of Oxford historians. This was viewed by the Oxford elite as a slap in the face, and the faculty of modern history there considered 'mak[ing] a unanimous protest against the appointment', the future Oxford Regius Professor of History C. H. Firth wrote to professor of History at the University of Manchester, T. F. Tout; '[e]verybody I have seen here feels strongly on this point'.[10] The editor of the *EHR*, R. Lane Poole, reported that 'Many of us were keen upon a public protest' and hopingly joked that Froude would 'give an inaugural lecture this term, draw a half year's salary, and resign in October'.[11]

Froude's 'Inaugural Lecture' began harmlessly enough as he observed just how different was the university from the time when he held his fellowship at

Exeter College and when the Oxford Movement was still in full swing. The university 'still stands', argued Froude, and 'it is full of animation and energy; but Keble and Newman are gone, and the system which produced such men is gone with them.' Indeed, Oxford was no longer an institution whose main function it was to train professional clerics. 'The celibate seclusion of college life has broken down', argued Froude, 'and ladies, the horror of the scholastics, have invaded the sacred precincts.' What was perhaps more important according to Froude was the rise of 'new schools' and 'new modes of teaching. Greek and Latin have lost their old monopoly. Modern languages are studied, and modern history, and modern philosophy and science.'[12] Oxford was, therefore, a very different place than the one Froude remembered when he was forced to resign his fellowship as a heretic. Much like Acton's appointment at Cambridge three years later, Froude's appointment as Regius Professor of Modern History at Oxford was a symbol of the vast changes that had transformed the ancient universities from primarily clerical training grounds to institutions devoted to higher learning.

Froude's subject matter would not be the transformation of the university system, however, but rather the way in which he believed history ought to be written. His argument certainly would have justified the concerns of proponents of the inductive science of history. Much like his Royal Institution lecture almost thirty years before, Froude explained that historical evidence was far too contingent for history ever to be regarded as a science. 'In history we have a record of things which happened once, or were said to have happened, but which, once passed, are gone forever.'[13] All facts, according to Froude, are merely 'supposed' because 'the writers on whom we depend were subject to the prejudices of their own times, and we who study them have prejudices of our own which appear in the form in which we re-tell their stories.' It is in this way, argued Froude, that historical knowledge is essentially 'mythic'. 'All history is mythic. Our knowledge of one another is mythic. Our knowledge of everything is mythic, for in every act of perception we contribute something of our own.' Historical knowledge, therefore, is entirely subjective and this is quite simply a fact that historians can no longer deny. 'We might as easily escape from our shadows ... We cannot escape our prejudices, which will and must guide us in the witnesses whom we trust.'[14]

Constructing an objective self of the kind advocated by Lord Acton was therefore impossible for Froude. Historians who no longer sought to escape from their shadows and who accept the subjective nature of historical knowledge, should look not to philosophers of history or proponents of historical methods but rather to the great dramatists, particularly that of Shakespeare. Froude believed that Shakespeare's plays offered a wonderful example of how the historian should approach his evidence, how it should be interpreted and disseminated, and how the stories of history should be told, in light of the subjective nature of all knowledge. Shakespeare 'does not moralise' about his characters, nor does he give the reader an 'opinion of his own'. Rather, Shakespeare 'gives you the men themselves to look at, to study, to reflect on, and (if you please) to form opinions about for yourself'. But Shakespeare does not moralize and he certainly 'does not invite *you* to draw lessons.' Such would almost certainly be impossible because the nature of his plays makes it difficult to put into words just what they are even supposed to mean. 'The more completely you have mastered these plays, the less you will be able to say what they have actually taught you.' This is because '[t]here is always something in the actions of men, and in men themselves, which escapes analysis'. And this is the nature of humanity that Shakespeare so brilliantly conveys, the actions and motivations that cannot be explained with words. As the historian is also concerned with representing 'the actions of men', '[i]f the historian would represent truly he must represent as the dramatist does.'[15]

Froude was adamant, however, that the historian as dramatist was not therefore a man of fiction. The only way the historian will be able to 'penetrate ... into the inner secrets of the past' is to 'study the original authorities. Go to the chronicles written by men who lived at the time and breathed the contemporary air. Drink at the fountain.' Froude directed his listeners to read contemporary letters where possible, '[r]ead what they say themselves', and go to state archives and examine government documents.[16] This was what he had done in researching his *History of England* and it was certainly not easy. 'I had to cut my way through a jungle, for no one had opened the road for me. I have been turned into rooms piled to the window-sill with bundles of dust-covered despatches, and told to make the best of it. Often I have found the sand glittering on the ink where it had been sprinkled when a page was turned.'[17] Indeed, Froude may have believed that facts were entirely subjective, but his works were built on a foundation of 'supposed' facts found in dusty archives and were given life by a man who embraced the artistic nuances of his craft.

Froude ended his lecture, naturally enough, by quoting Carlyle, his long-time mentor who believed that the 'history of mankind ... is the history of its great men.' This was a central belief that had guided Froude's own historical studies, one that he believed was still central to the discipline as a whole. What was history if not the study of great men and their actions? 'To find out these, clear the dirt from them and place them on their proper pedestals, is the function of the historian. He cannot have a nobler one.'

This call for a renewal of Carlyle's great man history would have surely reminded younger historians such as Firth, Tout and Poole just how old-fashioned was Froude's historical methodology despite his rather eloquent defence of an art of history that relies on extensive archival research. But the negative response to Froude's appointment was largely localized to a group of professional historians and was certainly not reflected in the periodical press. The *English Illustrated Magazine* did admit that Froude 'has not the temperament of a scientific historian', that he regarded 'history as only a branch of literature' and was 'an ardent hero-worshipper, a lover of paradox'. But these apparent objections against Froude's appointment, it was argued, were quite minor in the grand scheme of things. Instead we should rejoice at his appointment because of 'the stimulus that Mr. Froude's genius gave to the study of historical subjects, to rejoice that others, at the most impressionable period of their lives, will be brought within the range of his intellectual influence'.[18] If the *English Illustrated Magazine*'s support of Froude's appointment was not terribly surprising, Froude did receive some support from unlikely sources.

The *Saturday Review*, aka the 'Saturday Reviler', the weekly review that had spilled so much ink in defiling Froude's *History of England*, supported Froude's appointment to the Regius Chair, and not just grudgingly. 'The appointment of Mr. Froude in Mr. Freeman's place is humorous, unexpected, and satisfactory', explained the *Saturday Review*. The *Saturday* did not deny the fact that there was much 'rage' that accompanied the appointment, that 'Dryasdusts bewailed themselves, like the doleful creatures they are'. The *Saturday* also admitted that there were certainly significant problems with Froude's historical work, most notably his inaccuracies, but 'his inaccuracies have been quite sufficiently visited upon him' and there 'is no need to say any more on that side of the matter; something too much, indeed, may have been said as it was'. More strikingly still, the *Saturday* went on to say that 'Minute accuracy in detail is only the small game of history in any case'. We have clearly come a long way from the days when accuracy was the primary trait of a good historian and inaccuracy the black mark of a literary interloper, a point made quite often in the periodical now arguing that it is a minor blemish on the career of 'the greatest historical writer, if not the greatest prose writer of any kind, that England possesses'. The *Saturday Review* believed that Froude's appointment was 'the best which could have been made'. 'There has been no better appointment in Oxford since Mr. Freeman's own.'[19]

While the *Saturday* was willing to overlook its own past judgments in coming to terms with Froude's appointment many others were unable to do so. They would not have to endure Froude's appointment long, however, as he would die only two years into his tenure on 20 October 1894. We have seen (Chapter 7) that the *English Historical Review* became a wonderful space to memorialize the great names of history that passed away in the final decade of Queen Victoria's reign. James Anthony Froude would not be one of those historians. The *EHR* silently expressed its judgment on his career by not memorializing him in the way that was reserved for the other historians of his generation. A review did appear a year after he died and his death was at least mentioned. Jessop was much kinder in his review of Froude's *Life and Letters of Erasmus* than in his previous review but he admitted to having difficulty reviewing the piece honestly given Froude's recent death. Yet his previous judgments held true:

> Mr. Froude's literary faculty was transcendent; it placed him almost above criticism, it won for him a place in the very foremost rank of English prose writers; but among those who demand from the historian sobriety of judgment, severe accuracy of statement, and the subordination of the functions of advocate to those of the philosophic thinker – one capable of taking a calm survey of conflicting testimony and arriving at conclusions from large introduction unbiased by prejudice or passion – he never can be accepted as a trustworthy guide or a safe teacher to follow.[20]

Clearly some historians were still chanting the old scientific mantras and now using Froude's memory to continue to rehash old debates that others like the *Saturday Review* were finding no longer relevant.

A lengthy review of Froude's historical work appeared in the *Fortnightly Review* quite soon after his death and was written by H. A. L. Fisher (1865–1940) who seemed to be writing on behalf of the historical profession at large. Fisher was quite honest about Froude's capabilities. In particular, when it came to discharging his 'duties of his office' as Regius Professor, he did a wonderful job. Unlike his predecessors, Stubbs and Freeman, who lectured to empty benches and continually complained about their institutional duties, Froude clearly loved his work and his lectures were extremely popular. 'If the object of oral teaching be to stimulate rather than to satisfy curiosity', argued Fisher, 'to present subjects in an attractive manner, not to discharge the winnowings of a notebook, then Mr. Froude was the ideal lecturer'. But, of course, being popular and stimulating students were not the only tasks of the Regius Professor. Training the students in impartiality, accuracy, and in the proper usage of facts were considered important tasks too, and ones that Froude could not have been expected to do.

With that said, Fisher was also highly complimentary of the amount of research Froude would complete for each of his studies, even if his use and understanding of such research was suspect. 'His *History of England* was largely built upon unpublished material', argued Fisher, 'for which he had to consult the libraries of England, Spain, and of France'. Fisher was adamant that critics of Froude's work had to at the very least admit that Froude was one of the very first to make use of extensive archival material. 'Whatever may be said in depreciation of Mr. Froude's historical work, this must always be remembered, that he is one of the very small band of English historians who have based a comprehensive and artistic presentation of History upon palaeographical research.'[21]

The problem with Froude's historical work, argued Fisher, was not with the research or even the florid writing style, but with the faulty method that necessarily undermined what must have been extremely hard work. 'In a positive and scientific age ... [Froude] protested that there was no Science of History'.[22] Indeed, Froude not only believed that history could not be a science, he also believed that the connection between history and literature was 'not a deplorable accident of historical writing', as did the proponents of scientific history. He believed that literature was history's 'very soul and essence'. Not only that, but Froude seemed to suggest that 'history was not the most uncertain of the sciences: it was a branch of the dramatic arts, concerned with the revelation of truth'.[23] For Fisher, Froude significantly challenged the 'dignity of history' suggesting instead that it was merely a branch of artistic knowledge.

By denying history the possibility of becoming a science, Froude made history out to be a highly subjective discipline, not open to the rigors of objective scientific observation and experiment. Indeed, Froude contradicted just about everything scientific historians took as a given about their vocation. According to Fisher, Froude believed that:

> The past is gone for ever, and no magic can entice the knight from his tomb, or spell out the motives of the dead. We have no means of weighing our results in the laboratory; we cannot cross-question the spirits. ... The historical record is not only imperfect, it is also fraudulent. We who write history now, are at least two degrees removed from the truth. The contemporary authorities gave their own version of what they chose to see or hear, and we give our own version of what we choose to read in them.[24]

For scientific historians, Froude's mistakes were the inevitable result of adhering to history as a form of literature instead of as a form of science. History as literature would not only lessen history's standing within the professional disciplines, it would also make the knowledge claims made by historians entirely relativistic. In short, history could not exist as a discipline in Froude's methodology. Fisher, for his part, entirely disagreed with Froude's relativistic philosophy of history, arguing that the historical record may not be perfect but that 'the imperfections are often overrated. ... Even in the dimly lighted portions of history, the evidence generally points in one direction rather than in another, and we are bound to prefer the version which has more evidence in favour to the version which has less.'[25]

For Fisher, Froude's philosophy of history was not only erroneous but also, and more importantly, dangerous. Froude threatened to throw the judgment of history open to the whims of the public, rather than to the competently trained historian, whose disinterested analysis would only be seen as a relativistic version of what the historian himself chooses to write about. This was the reason why Fisher felt that Froude could not be considered a great historian: his denial of the science of history would deny the very existence of the discipline. The scientific historian was not asking for much, Fisher claimed, 'only that opinion on historical subjects be guided by the verdict of a competent expert, who knows the evidence'. Fisher argued that: 'So long as history is allowed to be concerned with truth', (suggesting that Froude was not concerned with truth), 'the true historian will prefer to be judged, not by the public, who may enjoy his style, but by the one or two specialists who can test his facts'.[26] Froude's artistic method of history threatened the very discipline of history itself.

As perhaps Fisher himself would have logically deduced, most of the periodical press was more positive about Froude's historical works and in particular about his artistic method, a method that, it was argued, actually saved history rather than attempted to kill it. An author for the *Speaker* triumphantly declared that Froude's histories will live – quite in contrast to his scientific contemporaries. In trying to answer the question as to whether or not Froude's works will live on past his death, the *Speaker* argued that his specifically fictional and political work would not. However, when it came to his historical work, it surely would. 'If [Froude] lives it will be as an historian', argued the *Speaker*, 'and if he lives as an historian it will be because he wrote history so that we cannot help reading it'. For this author it did not matter much that Froude was 'very inaccurate'. It was 'not fair to treat his inaccuracies as Freeman did. They involved no moral turpitude.' The author seemed to recognize that Freeman's main complaint was perhaps correct, that 'the picturesqueness, the air of reality, which carry one along, could not be obtained without the occasional use of imagination'. But where Freeman would argue that that is precisely why the imagination should be suppressed, the *Speaker* was quite willing to accept the odd inaccuracy for a history worth reading. 'Froude's inaccuracies are only the defects of his qualities', qualities that make history great. Froude will surely live on while the scientific historians will die, just like the historians of the previous century, of which 'only Gibbon will survive'.[27]

Similarly, the *Academy* argued that the judgments of professional historians will have little effect on the public's opinions of Froude's historical works, that 'the great reputation of Froude with the public will stand but little impaired'. Froude's scientific contemporaries seemed to misunderstand that 'the object of writing books is that they may be read; and in this respect Froude could afford to ignore the carpings of his critics'. A history book must be able to interest a reader in the same way that we would expect a professor to 'stimulate impressionable youth by his eloquence'. The fact that Froude was able to do both of these things, writing interesting books while providing 'the most effective lectures that have been heard at [Oxford] University since the time of Matthew Arnold and Ruskin', places him well above the scientific historians who claimed that their first duty was to please historical peers.[28]

Given the *Saturday Review*'s reversal of opinion on Froude, it is not surprising that he received a warm remembrance in the weekly's pages as well. Interestingly, however, the *Saturday* did not try to judge and compare Froude's career with that of other historians; he was rather considered alongside other literary heavyweights of the Victorian period, such names as the already deceased Alfred Tennyson and Robert Browning, John Henry Newman and Matthew Arnold. Before Froude's death one could only cite he and Ruskin as the remaining 'captains of prose and verse' and now there 'is no one but Mr. Ruskin ... of the first class of veterans' the *Saturday* lamented after recording Froude's death.[29] The *Saturday* in particular highlighted Froude's 'middle work', 'of the ordinary tissue of his history and his essays' as deserving 'the highest of praise'. The *Saturday* expressed slight disbelief at 'How such a writer, with such a love for historical writing, could lay himself open as Mr. Froude did to the attacks of men who were unworthy to loose his shoe-latchets in point of style, genius, patriotism and even real historic grasp of the general kind', attacks that the *Saturday* was only too happy to publish in its pages. It was perhaps because 'Froude overran half the fields of literature and history with a step as careless as it was confident', that he seemed not to care when he ruffled feathers or caused controversy or attempted to overturn well-established historical orthodoxy. That he certainly made mistakes in doing so does not sully the fact that '[h]e wrote some of the best English of his time' and that 'he loved England heart and soul'.[30]

The general public's embrace of Froude's literary and historical career suggests that the boundary work directed at the original sufferer of Froude's disease only succeeded within the historical profession, where Froude was more explicitly excluded. The art of history remained a powerful method as far as the general public was concerned, and the profession of history in Britain would once again, though rather reluctantly, have to make room for artistic and literary methods to exist alongside scientific ones. This transformation likely occurred over time and there are certainly subtle signs that the orthodox science of history began to break down by the end of the century, particularly in the obituaries analysed in the previous chapter. However, a much more explicit sign appeared in response to the inaugural lecture of Acton's successor, J. B. Bury (1861–1927).

Bury had been one of the bright lights of the second generation of professional historians. He was primarily a classical scholar and he wrote fairly well-received books on the Roman Empire that were published in the 1890s. He was also a regular contributor to the *English Historical Review*, publishing most notably an article in the inaugural volume. He was one of the first historians Acton considered to write something for the *Cambridge Modern History*. There was little surprise when he was chosen to succeed Acton in 1902. His inaugural lecture of 1903 is illustrative of just how influenced he was by the first generation of scientific historians and how badly he wanted to continue to support their legacy.

He began by admitting the enhanced 'terrors' of taking up the chair which had been filled by the great Lord Acton. Bury felt that he should avoid using the forum to put forward any new argument about the nature of history, or point towards any great new subject matter or methodology; instead he believed it was his duty 'to pay a sort of solemn tribute to the dignity and authority' of history by 'reciting some of her claims and her laws, or by reviewing the measures of her dominion'. Bury observed that the discipline had gone through a profound transformation within the short space of three generations. History 'began to forsake her old irresponsible ways and prepared to enter into her kingdom'.[31] He argued that the transformation was not quite complete because there were still those who refuse to admit that history is a science.[32]

Bury proceeded to explain how history became a science over the course of the previous three generations. He spoke of history's development into an autonomous discipline of knowledge because of the adoption of the German critical research method. He discussed the boundaries that were constructed to separate history from literature and he rejoiced that history had finally 'begun to enter into closer relations with the sciences which deal objectively with the facts of the universe'. The lecture sounded very much like Acton's a few years before and even Stubbs's now almost forty-year-old inaugural. Bury concluded his lecture by repeating a phrase he mentioned at the outset: that history was 'simply a science, no less and no more'.[33] On the basis of his lecture, it appears as if the discipline of history was still quite wedded to the scientific methodology promoted by the likes of Acton and Stubbs.

However, if Bury's lecture is illustrative of a continued belief in the science of history, it is also illustrative of the fact that this was no longer the consensus view. Herbert Paul, writing for the *Speaker*, was absolutely dumbfounded that such a simplistic argument could be pronounced from the Regius chair. 'To many people, and not those who have studied history with the least attention', Paul argued, the statement that history is only a science 'will seem partly false and partly unmeaning'.[34] Paul went on to say that the many historians Bury cited as adhering to history as only a science, relied on much more than science to complete their works. The German historian Niebuhr, for instance, 'carried the imaginative reconstruction of history ... beyond what all sober judgment would allow'. History without literature 'is not complete, for it lacks the touch of the true historian'. Paul was glad, however, that most historians no longer openly adhered to such an unattractive methodology. 'Happily it is not attractive enough to discourage the composition of such excellent and truly historical books as Mr. Trevelyan's *England in the Age of Wycliffe* [1899]'.[35]

The young G. M. Trevelyan (1876–1962) was becoming the new face of the historical profession in England. The great-nephew of Macaulay, he wrote in a similar literary style and his primary audience was the general reader, but he also based his studies on painstaking archival research. Lord Acton was impressed with his early historical work as a student of Cambridge and believed he had uncovered documents no one had ever seen before in his 'Prize Essay on the last years of Edward III'.[36] Trevelyan would also respond to Bury's lecture and his criticisms suggest that he was more than an observant student while he read history at Cambridge in the 1890s.

In responding to Bury's lecture, Trevelyan insisted that history was not simply a science but was also a necessarily artistic endeavour. That history was no longer a branch of literature should not deter the historian from enlisting the artist's sensibility. History was a science in its methods of collecting and evaluating evidence, but in the representation of such evidence, Trevelyan argued, history was just as much art as science. In order to buttress his claim, Trevelyan pointed out that '[n]early all the great leaders of English scientific history – Seeley, Creighton, Gardiner, Freeman – were literary men as well as scientists'. Trevelyan believed that such historians argued so violently on behalf of scientific history because 'it was in order to win for the science a recognition yet more complete than it then enjoyed'. There was no longer any reason to denounce history's artistic merits, Trevelyan claimed, because science had already been established as a part of the historian's method. In other words, Trevelyan believed that Bury was fighting a battle that had already been won.[37] History is indeed a science. But it also a form of art.

Bury was soon after in 1909 overridden by the Board of History at Cambridge in his attempt to severely limit the study of history to a narrow examination of particular facts under a supposedly rigorous scientific methodology. Apparently siding with Trevelyan, the Board ruled that 'a correct general knowledge' was more important 'than minute acquaintance with details'. Examiners were therefore encouraged to give credit for both 'style and method'. Appropriately, Trevelyan replaced Bury as Regius Professor of Modern History at Cambridge in 1927 (just as Froude replaced Freeman at Oxford) and he repeated much of his criticisms of scientific history in his inaugural lecture.[38]

If the battle between the science and art of history seemed to end in a draw by the time of Trevelyan's famous response to Bury in 1903, the twentieth century certainly continued this battle, drawing subtly on arguments that seemed to be largely exhausted in the nineteenth century. Froude's more relativist approach became quite popular after the First World War, particularly in the United States, only to be driven underground during and after the Second World War by renewed calls to make history a science and truly objective. Covering Lawyers, Annales historians and quantifiers of various stripes all attempted to bring elements of Buckle's positivist dream back to life without naming the source. Meanwhile the dryasdust fact-grubbers originally denounced by Thomas Carlyle and Walter Scott only to be made into a reality by Freeman and others kept plugging away at their specialized studies. Criticism of their work from outside the ivory tower continued as well, most notably by Kingsley Amis who warned of the fanatical boredom produced by academic historians in *Lucky Jim* (1954). In the 1970s and 1980s, historians began to yet again pay more attention to their writing style – the so-called linguistic turn – this time proposing no longer to hide behind their reality effects and accept the contrived nature of their procedures.[39] Hayden White's call, in particular, for historians to take advantage of modernist literary techniques in order to get even closer to the sublime reality of the past reads very much like Froude's own embrace of the historian as dramatist.[40] Presently, most historians would likely agree with Froude in saying that history is *not* a science and that historical knowledge is *not* objective. And yet we professional historians still clearly fear that history written for the general reader by a non-specialist and continue to find new and disturbing ways to police our professional boundaries. Instead of criticizing works as suffering from Froude's disease we denounce them as being a product of, for instance, the 'Sobel Effect', a publisher's popular template for making the history of science fascinating, and laugh at the inevitable 'howlers' that are produced.[41] Bound up within these debates and boundary disputes is a fundamental question about the nature of historical analysis: is it a science or a form of art? Perhaps that is simply a battle which – according to Buckle – 'will never be ended'.

NOTES

Introduction: That Never-Ending Battle

1. Henry Thomas Buckle to Maria Grey, 23 February 1853, H. T. Buckle Collection, University of Illinois at Urbana-Champaign (hereafter cited as Buckle Collection), MS 66, vol. 10, p. 10; and A. H. Huth, *The Life and Writings of Henry Thomas Buckle*, 2 vols (London: Sampson Low, Marston, Searle Rivington, 1880), vol. 1, p. 81.
2. Buckle to Grey, 18 September 1854, Buckle Collection, MS 66, vol. 10, p. 13; and Huth, *Life and Letters of Henry Thomas Buckle*, vol. 1, p. 82.
3. On the diverse forms of scientific history-writing in nineteenth-century Europe (including Buckle's) see E. Fuchs, 'Contemporary Alternatives to German Historicism in the Nineteenth Century', in S. Macintyre, J. Maiguashca and A. Pók (eds), *The Oxford History of Historical Writing*, vol. 4: *1800–1945* (Oxford: Oxford University Press, forthcoming, 2011).
4. H. T. Buckle, *History of Civilization in England*, 2 vols (1857, 1861; New York: D. Appleton and Company, 1860, 1861), vol. 1, pp. 166–7.
5. C. Dickens, *Hard Times* (London, 1854), ch. 1.
6. H. T. Buckle, 'The Influence of Women on the Progress of Knowledge', in *Miscellaneous and Posthumous Works of Henry Thomas Buckle*, ed. H. Taylor, 3 vols (London: Longmans, Green, and Co., 1872), vol. 1, pp. 1–19, on pp. 9–10.
7. Buckle was certainly not alone in wanting to maintain a balance between science and art and this was actually a prominent area of concern among those who sought to follow Comte and apply the scientific method to society like Buckle and John Stuart Mill and other figures often associated with the founding of sociology. See W. Lepenies, *Between Literature and Science: The Rise of Sociology*, trans. R. J. Hollingdale (Cambridge: Cambridge University Press, 1988), esp. ch. 3.
8. On the rejection of Buckle by historians see esp. C. Parker, 'English Historians and the Opposition to Positivism', *History and Theory*, 22:2 (1983), pp. 120–45.
9. Walter Scott cheekily dedicated some of his novels to the fictitious J. Dryasdust for supplying him with the necessary dry historical details. On Romantic history see S. Bann, *Romanticism and the Rise of History* (New York: Twayne Publishers, 1995); M. S. Philips, *Society and Sentiment: Genres of Historical Writing in Britain, 1740–1820* (Princeton, NJ: Princeton University Press, 2000); A. Rigney, *Imperfect Histories: The Elusive Past and the Legacy of Romantic Historicism* (Ithaca, NY: Cornell University Press, 2001); S. Berger, 'The Invention of European National Traditions in European Romanticism' in Macintyre, Maiguashca and Pók (eds), *The Oxford History of Historical Writing*, vol. 4;

and J. Tollebeek, 'Seeing the Past with the Mind's Eye: The Consecration of the Romantic Historian', *Clio*, 29:2 (2000), pp. 167–91.
10. On the revolution in printing and literacy see R. D. Altick, *The English Common Reader: A Social History of the Mass Reading Public, 1800–1900*, 2nd edn (Columbus, OH: Ohio State University, 1998); and as it relates to science publishing see J. Topham, 'Scientific Publishing and the Reading of Science in Nineteenth-Century Britain: A Historiographical Survey and Guide to Sources', *Studies in the History and Philosophy of Science*, 31:4 (2000), pp. 559–612.
11. Rigney, *Imperfect Histories*. For a general analysis of the general blurring of the boundaries between history and literature that does focus particular attention on the early nineteenth century see B. Southgate, *History Meets Fiction* (Harlow: Pearson, 2009).
12. [T. Carlyle], 'Thoughts on History', *Fraser's Magazine*, 10:2 (November 1830), pp. 413–18 on p. 413.
13. On Anglicanism and Victorian historians see J. P. von Arx, *Progress and Pessimism: Religion, Politics, and History in Late Nineteenth-Century Britain* (Cambridge, MA: Harvard University Press, 1985); and D. Forbes, *The Liberal Anglican Idea of History* (Cambridge: Cambridge University Press, 1953).
14. See, in particular, L. von Ranke, 'Preface to the First Edition of Histories of the Latin and Germanic Nations (October 1824)', in *The Theory and Practice of History*, ed. G. G. Iggers and K. Von Moltke (Indianapolis, IN: Bobbs-Merrill, 1973), p. 137. In the introduction to this work, Iggers points out that Ranke's dictum should likely read 'how things, essentially, happened' and that this mistranslation has led to much confusion as to Ranke's precise meaning. This was particularly relevant in the American context according to P. Novick's *That Noble Dream: The 'Objectivity Question' and the American Historical Profession* (Cambridge: Cambridge University Press, 1988), ch. 1. Recent studies suggest, however, that too much has been made of the mistranslation: E. K-Ma Cheng, 'Exceptional History? The Origins of Historiography in the United States', *History and Theory*, 47 (2008), pp. 200–28; and D. Ross, 'On the Misunderstanding of Ranke and the Origins of the Historical Profession in America', in G. G. Iggers and J. M. Powell (eds), *Leopold von Ranke and the Shaping of the Historical Discipline* (Syracuse, NY: Syracuse University Press, 1990), pp. 154–69. See also S. Bann, *The Clothing of Clio: A Study of the Representation of History in Nineteenth-Century Britain and France* (Cambridge: Cambridge University Press, 1984), for French and British variations of the Rankean dictum.
15. On the establishment of the disinterested Baconian identity in English science see S. Shapin, *A Social History of Truth: Civility and Science in Seventeenth-Century England* (Chicago, IL: University of Chicago Press, 1994). On Robert Boyle as the primary proponent and practitioner of this method see S. Shapin and S. Schaffer, *Leviathan and the Air-Pump: Hobbes, Boyle, and the Experimental Life* (Chicago, IL: University of Chicago Press, 1985).
16. R. Yeo, 'An Idol of the Market-Place: Baconianism in Nineteenth-Century Britain', *History of Science*, 23 (1985), pp. 251–98. For the centrality of Baconianism in literature see J. Smith, *Fact and Feeling: Baconian Science and the Nineteenth-Century Literary Imagination* (Madison, WI: University of Wisconsin Press, 1994).
17. J. A. Froude, 'The Science of History: A Lecture Delivered at the Royal Institution, February 5, 1864', in *Short Studies on Great Subjects*, 4 vols (London: Longmans, Green, and Co., 1894), pp. 1–25; and C. Kingsley, 'The Limits of Exact Science as Applied to History', in *The Roman and the Teuton* (London: Macmillan and Co., 1864), appendix.

18. W. Stubbs, 'Inaugural. (Feb 7, 1867)', in *Seventeen Lectures on the Study of Medieval and Modern History and Kindred Subjects* (New York: Howard Fertig, 1967), p. 11.
19. Lord Acton, 'Mr. Goldwin Smith's Irish History', *Rambler*, 6 (January 1862), pp. 190–229, reprinted in *Selected Writings of Lord Acton*, vol. 2: *Essays in the Study and Writing of History*, ed. J. R. Fears (Indianapolis, IN: Liberty Fund, 1986), pp. 67–97 on p. 69.
20. [E. A. Freeman], 'The Art of History-Making', *Saturday Review*, 17 November 1855, pp. 52–4 on p. 52.
21. The authoritative work on popularizers of science is B. Lightman, *Victorian Popularizers of Science: Designing Nature for New Audiences* (Chicago, IL: University of Chicago Press, 2007). For the explicit use of story-telling techniques by popularizers of science in particular see Lightman, 'The Story of Nature: Victorian Popularizers and Scientific Narrative', *Victorian Review*, 25:2 (1999), pp. 1–29.
22. J. Second, *Victorian Sensation: The Extraordinary Publication, Reception, and Secret Authorship of* Vestiges of the Natural History of Creation (Chicago, IL: University of Chicago Press, 2000); and R. Yeo, 'Science and Intellectual Authority in Mid-Nineteenth-Century Britain: Robert Chambers and "Vestiges of the Natural History of Creation"', *Victorian Studies*, 28 (1984), pp. 5–31.
23. F. M. Turner, *Contesting Cultural Authority: Essays in Victorian Intellectual Life* (Cambridge: Cambridge University Press, 1993); A. Desmond, *The Politics of Evolution: Morphology, Medicine, and Reform in Radical London* (Chicago, IL: University of Chicago Press, 1989); B. Lightman, *Evolutionary Naturalism in Victorian Britain: The 'Darwinians' and Their Critics* (Burlington, Vt.: Ashgate, 2009), esp. ch. 1; Lightman (ed.), *Victorian Science in Context* (Chicago, IL: University of Chicago Press); C. Smith, *The Science of Energy: A Cultural History of Energy Physics in Victorian Britain* (Chicago, IL: University of Chicago Press, 1998); and M. J. S. Rudwick, *The Great Devonian Controversy: The Shaping of Scientific Knowledge among Gentlemanly Specialists* (Chicago, IL: University of Chicago Press, 1985).
24. On boundary work of a professionalizing science see T. F. Gieryn, *Cultural Boundaries of Science: Credibility on the Line* (Chicago, IL: University of Chicago Press, 1999).
25. J. R. Seeley, 'History and Politics I', *Macmillan's Magazine*, 40 (1879), pp. 289–99.
26. E. A. Freeman, 'Lecture II: The Difficulties of Historical Study', in *The Methods of Historical Study* (London: The Macmillan Company, 1886), pp. 99–100.
27. See I. Hesketh, 'Diagnosing Froude's Disease: The Discipline of History in Late-Victorian Britain', *History and Theory*, 47 (2008), pp. 373–97.
28. L. Daston and P. Galison, *Objectivity* (New York: Zone Books, 2007).
29. T. M. Porter, 'The Objective Self', *Victorian Studies*, 50 (2008), pp. 641–7.
30. 'The Literary Week', *Academy*, 18 May 1901, p. 415.
31. Huxley is a great example of the new 'man of science' who appears post-1859, straddling the boundaries between professional scientist and public intellectual. See in particular Adrian Desmond's authoritative biography *Huxley: From Devil's Disciple to Evolution's High Priest* (London: Penguin Books, 1997); but also see Paul White, *Thomas Huxley: Making the 'Man of Science'* (Cambridge: Cambridge University Press, 2003) for Huxley's strategies of self-presentation. For the linkage between scientific identity and practice in other central Victorian 'men of science' see also J. Endersby, *Imperial Nature: Joseph Hooker and the Practices of Victorian Science* (Chicago, IL: University of Chicago Press, 2008); and (despite the overuse of the still anachronistic term 'scientist') J. Meadows, *The Victorian Scientist: The Growth of a Profession* (London: The British Library, 2004).

32. L. Howsam, 'Academic Discipline or Literary Genre? The Establishment of Boundaries in Historical Writing', *Victorian Literature and Culture* (2004), pp. 525–45; Howsam, 'Imperial Publishers and the Idea of Colonial History, 1870–1916', *History of Intellectual Culture* (2006), pp. 1–30; and Howsam, *Past into Print: The Publishing of History in Britain, 1850–1950* (London: The British Library; Toronto: University of Toronto Press, 2009). See also R. Mitchell, *Picturing the Past: English History in Text and Image, 1830–1870* (Oxford: Clarendon Press, 2000).
33. On twentieth-century historical writing in Britain see in particular R. N. Soffer, *History, Historians and Conservatism in Britain and America: The Great War to Thatcher and Reagan* (Oxford: Oxford University Press, 2009); P. Burke, *History and Historians in the Twentieth Century* (Oxford: Oxford University Press, 2002); C. Parker, *The English Historical Tradition since 1850*; M. Bentley, *Modernizing England's Past: English Historiography in the Age of Modernism, 1870–1970* (Cambridge: Cambridge University Press, 2005); and Bentley, 'British Historical Writing', in A. Scheider and D. Woolf (eds), *The Oxford History of Historical Writing*, vol. 5: *Historical Writing since 1945* (Oxford: Oxford University Press, forthcoming, 2011).
34. Heyck, *The Transformation of Intellectual Life in Victorian England*, ch. 5; D. S. Goldstein: 'The Professionalization of History in Britain in the Late Nineteenth and Early Twentieth Centuries', *Storia della Storiographia*, 3 (1983), pp. 3–25; J. Kenyon, *The History Men: The Historical Profession in England since the Renaissance* (London: Weidenfeld and Nicolson, 1983), ch. 5; and Parker, 'English Historians and the Opposition to Positivism'.
35. For general studies see J. W. Burrow, *A Liberal Descent: Victorian Historians and the English Past* (Cambridge: Cambridge University Press, 1981); A. Dwight Culler, *The Victorian Mirror of History* (New Haven, CT: Yale University Press, 1985); and R. Jann, *The Art and Science of Victorian History* (Columbus, OH: Ohio State University Press, 1985). The best modern studies of individual historians are: D. Wormell, *Sir John Seeley and the Uses of History* (Cambridge: Cambridge University Press, 1980); J. Markus, *J. Anthony Froude: The Last Undiscovered Great Victorian* (New York: Scribner, 2005); and R. Hill, *Lord Acton* (New Haven/New York: Yale University Press, 2000).
36. P. R. H. Slee, *Learning and a Liberal Education: The Study of Modern History in the Universities of Oxford, Cambridge, and Manchester, 1840–1914* (Manchester: Manchester University Press, 1986); P. Levine, *The Amateur and the Professional: Antiquarians, Historians and Archaeologists in Victorian England, 1838–1886* (Cambridge: Cambridge University Press, 1986); and R. N. Soffer, *Discipline and Power: The University, History, and the Making of an English Elite, 1870–1930* (Stanford, CA: Stanford University Press, 1994).
37. The classic study of Whig historiography is H. Butterfield, *The Whig Interpretation of History* (1931; New York: Norton, 1965).
38. See, for instance, Bentley, *Modernizing England's Past*; Bentley, 'Shape and Pattern in British Historical Writing, 1815–1945', in Macintyre, Maiguashca and Pók (eds), *The Oxford History of Historical Writing*, vol. 4; P. B. M. Blass, *Continuity and Anachronism: Parliamentary and Constitutional Development in Whig Historiography and in the Anti-Whig Reaction between 1890 and 1930* (The Hague: Nijhoff, 1978); Jann, *The Art and Science of Victorian History*; and Burrow, *A Liberal Descent*.
39. S. J. Gould, *Wonderful Life: The Burgess Shale and the Nature of History* (New York: Norton, 1989), p. 51. I am indebted to J. Beatty, 'Replaying Life's Tape', *The Journal of Philosophy*, 103 (2006), pp. 319–36, for the specifically historical implications of Gould's thought experiment.

40. See, in particular, H. White, *The Content of the Form: Narrative Discourse and Historical Representation* (Baltimore, MD: The Johns Hopkins University Press, 1987) (on the necessity of literary modes of representation); D. Lowenthal, *The Past is a Foreign Country* (Cambridge: Cambridge University Press, 1985) (on the incomprehensibility of the past); M. Frayn, *Copenhagen* (1998; New York: Anchor Books, 2000) (on uncertainty); and A. Confino, 'Narrative Form and Historical Sensation: On Saul Friedländer's *The Years of Extermination*', *History and Theory*, 48 (2009), pp. 199–219 (on descriptive illustrations vs. causal analysis).
41. It has become somewhat fashionable of late to draw explicit parallels between 'romantic historicism' and more contemporary considerations of historical indeterminacy and other so-called postmodern reflections on the nature of the past and historical writing. See, in particular, Rigney, *Imperfect Histories*; and Southgate, *History Meets Fiction*. Recent historians of the *longue durée* of historical writing, however, would suggest that these ideas originate well before the Romantic period and traces of them can certainly be found at least as early as ancient Greece, appearing and re-appearing in different forms at different times and places since then. See, in particular, D. Woolf, *A Global History of History* (Cambridge: Cambridge University Press, forthcoming, 2011); and D. R. Kelley, *The Faces of History: Historical Inquiry from Herodotus to Herder* (New Haven, CT: Yale University Press, 1998)

1 The Enlarging Horizon

1. Huth, *The Life and Writings of Henry Thomas Buckle*, vol. 1, p. 255. For Buckle's attempts to get tickets for his friends see Buckle to Theodore Parker, 10 March 1858, Buckle Collection, MS 66 vol. 10, p. 186; vol. 1, p. 4; and Buckle to Mrs Grey, 14 March 1858, Buckle Collection, MS 66 vol. 10, p. 138.
2. Quoted in Huth, *The Life and Writings of Henry Thomas Buckle*, vol. 1, p. 255.
3. Dr Whewell to Mrs Austin, 1 April 1858, in *Three Generations of English Women: Memoirs and Correspondence of Mrs. John Taylor, Mrs. Sarah Austin, and Lady Duff Gordon*, ed. J. Ross, 3 vols (London: John Murray, 1888), vol. 2, p. 56. On Whewell's formative role in early Victorian science see R. Yeo, *Defining Science: William Whewell, Natural Knowledge and Public Debate in Early Victorian Britain* (Cambridge: Cambridge University Press, 1993).
4. On Buckle's popularity see in particular B. Semmel, 'H. T. Buckle: The Liberal Faith and the Science of History', *British Journal of Sociology*, 27 (1976), pp. 370–86 on pp. 372–3; and T. M. Porter, 'Buckle, Henry Thomas', in B. Lightman (ed.), *The Dictionary of Nineteenth-Century British Scientists*, vol. 1: *A–C* (Bristol: Thoemmes Continuum, 2004), p. 334.
5. L. Stephen, 'An Attempted Philosophy of History', *Fortnightly Review*, 27 (1880), pp. 672–95 on p. 672.
6. For the events surrounding Buckle's death see Buckle Collection, University of Illinois at Urbana-Champaign (hereafter cited as Buckle Collection), MS 66, vol. 6; Huth, *The Life and Writings of Henry Thomas Buckle*, vol. 2, appendix; and G. St. Aubyn, *A Victorian Eminence: The Life and Works of Henry Thomas Buckle* (London: Barrie, 1958), ch. 4 and app. 2.
7. Buckle to Maria Grey, 30 June 1856, Buckle Collection, MS 66, vol. 10, p. 67.
8. Huth, *The Life and Writings of Henry Thomas Buckle*, vol. 1, pp. 20–1, 30.

9. Buckle to the Lord Kintore, 23 February 1853, Buckle Collection, MS 66, vol. 10, p. 4; and Huth, *The Life and Writings of Henry Thomas Buckle*, vol. 1, pp. 63–4.
10. [L. Stephen], 'Buckle, Henry Thomas (1821–1862)', in L. Stephen and S. Lee (eds), *The Dictionary of National Biography*, vol. 3 (1885; Oxford: Oxford University Press, 1917), pp. 208–12 on p. 209.
11. Quoted in St. Aubyn, *A Victorian Eminence*, p. 12.
12. 'I do not like reading in public libraries, and I purchase nearly all the books which I use. I have at present about 20 000 volumes.' Buckle to Theodore Parker, 9 July 1859, Buckle Collection, MS 66, vol. 10, p. 150.
13. The previous two paragraphs are based on St. Aubyn, *A Victorian Eminence*, pp. 1–12.
14. B. Powell, 'On the Study of the Evidences of Christianity', in *Essays and Reviews: The 1860 Text and Its Reading*, ed. V. Shea and W. Whitla (Charlottesville/London: University of Virginia, 2000). On the *Essays and Reviews* controversy see J. L. Altholz, *Anatomy of a Controversy: The Debate over Essays and Reviews, 1860–1864* (Aldershot: Scolar Press, 1994). On Baden Powell see P. Corsi, *Science and Religion: Baden Powell and the Anglican Debate, 1800–1860* (Cambridge: Cambridge University Press, 1988), esp. p. 205 for the friendship between Powell and Buckle (a subject that is still unfortunately understudied despite Corsi's early consideration).
15. Buckle to Maria Grey, 25 August 1861, Buckle Collection, MS 66, vol. 10, p. 282.
16. Buckle to Emily Shirreff, 23 August 1855, Buckle Collection, MS 66, vol. 10, p. 43. See also Buckle to Shirreff, 5 July 1855, MS 66 vol. 10, p. 37.
17. Buckle to Mrs Woodhead, 17 January 1860, Buckle Collection, MS 66, vol. 10, pp. 245–6.
18. For Comte's influence see T. R. Wright, *The Religion of Humanity: The Impact of Comtean Positivism on Victorian Britain* (Cambridge: Cambridge University Press, 1986); C. Kent, *Brains in Numbers: Elitism, Comtism, and Democracy in Mid-Victorian England* (Toronto: University of Toronto Press, 1978); W. M. Simon, *European Positivism in the Nineteenth Century: An Essay in Intellectual History* (Port Washington, NY: Kennijat Press, 1972); and J. M. Murphy, *Positivism in England: The Reception of Comte's Doctrines, 1840–1870* (New York: Columbia University, 1968).
19. Buckle, *History of Civilization in England*, vol. 1, pp. 427 n. 242, 4 n. 1. See also ibid., pp. 181 n. 19, 43 n. 33. On Buckle's embrace of a 'positivist' approach to the study of history see, in particular, E. Fuchs, *Henry Thomas Buckle: Geschichtschreibung und Positivismus in England und Deutschland* (Leipzig: Leipziger Universitätsverlag, 1994); and Fuchs, 'Contemporary Alternatives to German Historicism in the Nineteenth Century'.
20. A. Comte, 'The Positive Philosophy and the Study of Society', in P. Gardiner (ed.), *Theories of History* (New York: The Free Press, 1959), pp. 75–9.
21. Buckle, *History of Civilization in England*, vol. 1, p. 5.
22. Ibid., vol. 1, p. 7.
23. Ibid., vol. 1, pp. 14–15.
24. Ibid., vol. 1, ch. 2.
25. Ibid., vol. 1, p. 168.
26. Ibid., vol. 1, p. 210.
27. Ibid., vol. 1, p. 166.
28. Huth, *The Life and Writings of Henry Thomas Buckle*, vol. 1, pp. 239–42.

29. H. T. Buckle, 'Mill on Liberty', *Fraser's Magazine*, 59 (May 1859), pp. 509–42 on p. 511; also published in *Miscellaneous and Posthumous Works of Henry Thomas Buckle*, ed. Taylor, vol. 1, pp. 29–70, on p. 24.
30. I. Hacking, *The Taming of Chance* (Cambridge: Cambridge University Press, 1990), p. 105.
31. For the influence of statistics on Buckle's thinking see in particular H. Small, 'Chances Are: Henry Buckle, Thomas Hardy, and the Individual at Risk', in Small and T. Tate (eds), *Literature, Science, Psychoanalysis, 1830–1970* (Oxford: Oxford University Press, 2005), pp. 64–85; and T. M. Porter, *The Rise of Statistical Thinking, 1820–1900* (Princeton, NJ: Princeton University Press, 1986), pp. 60–5.
32. Buckle, *History of Civilization in England*, vol. 1, p. 20.
33. Ibid., vol. 1, pp. 19, 23, 24, 25.
34. Ibid., vol. 1, p. 128.
35. Ibid., vol. 1, pp. 128–9.
36. Buckle to Vice-Chancellor Wood, 31 October 1857, Buckle Collection, MS 66, vol. 10, pp. 117–19; and Huth, *The Life and Writings of Henry Thomas Buckle*, vol. 1, p. 126.
37. Stephen, 'An Attempted Philosophy of History', p. 672.
38. On the opposition of historians to Buckle's project see Parker, 'English Historians and the Opposition to Positivism'.
39. Buckle, *History of Civilization in England*, vol. 1, p. 163.
40. Buckle, 'Mill on Liberty', p. 510; and *Miscellaneous and Posthumous Works of Henry Thomas Buckle*, ed. Taylor, vol. 1, pp. 21–2.
41. Buckle, *History of Civilization in England*, vol. 1, p. 177.
42. [T. B. Macaulay], 'Lord Bacon', *Edinburgh Review*, 65 (1837), pp. 3–104; and D. Brewster, *The Life of Sir Isaac Newton* (1831; New York: Harper, 1840), esp. ch. 19. See also Yeo, 'An Idol of the Market-Place'; and Smith, *Fact and Feeling*, ch. 1.
43. Buckle to Parker, 9 December 1858, Buckle Collection, MS 66 vol. 10, pp. 168–9.
44. J. S. Mill, *A System of Logic* (1843) 8th edn (London: Longmans, 1967), book 5, ch. 11.
45. G. Capel to Mill, 30 May 1868, Buckle Collection, MS 66 vol. 4, p. 7; Mill to J. Buckle, 27 November 1868, Buckle Collection, MS 66 vol. 4, p. 14; and Miss Rogers to H. Huth, 13 October 1866, Buckle Collection, MS 66 vol. 6, p. 12. The posthumous collection would be published in 1872 by Longmans in three volumes as *Miscellaneous and Posthumous Works of Henry Thomas Buckle*.
46. J. S. Mill to F. B. Chapman, 24 February 1863, Buckle Collection, MS 66 vol. 4, p. 15. For Mill's specific criticisms see J. M. Robertson, *Buckle and His Critics: A Study in Sociology* (London: Swan Sonnenschein, 1895), pp. 269–79.
47. Buckle to H. Huth, 2 February 1862, Buckle Collection, MS 66, vol. 10, p. 312; and Huth, *Life and Writings of Henry Thomas Buckle*, vol. 2, pp. 157–8. Buckle gave a similar rationale to Maria Grey who, in attempting to benefit from Buckle's newfound celebrity, was hoping to count on his support for what seemed to be the creation of a school of medicine for women. 'I cannot openly countenance what I believe to be an extremely bold experiment, of which the evil (to my mind at least) is greater than the good.' He was, to be fair, more concerned about the general reaction of men who would as a result find themselves in a position of 'envy' and would therefore react accordingly: 'if the stronger sex should envy the weaker, it must happen that the weaker will go to the wall'. This would be another example of the reaction to reform proving more detrimental than the original social evil that was sought to be remedied. Buckle to Maria Grey, 18 March 1859, Buckle Collection, MS 66 vol. 10, pp. 183–5; and Huth, *Life and Writings of Henry Thomas*

48. [J. S. Mill], 'Bentham', *The London and Westminster Review*, 31 (August 1838), pp. 467–506; and [Mill], 'Coleridge', *The London and Westminster Review*, 33 (March 1840), pp. 257–302. See also W. Lepenies, *Between Literature and Science: The Rise of Sociology*, trans. R. J. Hollingdale (Cambridge: Cambridge, 1988), pp. 102–6.
49. I have been greatly informed by L. J. Snyder, *Reforming Philosophy: A Victorian Debate on Science and Society* (Chicago, IL: University of Chicago Press, 2006), who convincingly argues that Mill rejected the intuitionist epistemology because it would necessarily, in his mind at least, assure proponents 'that what they believed deeply must be true' (p. 96). Snyder further argues that Mill's *System of Logic* must be understood as a reformist text that was specifically directed against the intuitionist method offered by Whewell.
50. For Buckle's various notes on Whewell's work see *Miscellaneous and Posthumous Works of Henry Thomas Buckle*, ed. Taylor, vol. 1, pp. 162–3, 167, 169, 220, 283, 289, 398, 422, 505–7, 510–11, 530.
51. For Buckle's criticisms of Whewell's Bridgewater Treatise and other remnants concerning his natural theology see Buckle, *History of Civilization in England*, vol. 1, pp. 270 n. 78, 182 n. 22, 427 n. 242. On Whewell's argument concerning induction leading to Divine Truth see Snyder, *Reforming Philosophy*, pp. 23–4.
52. Ibid., vol. 2, p. 395.
53. Ibid., vol. 2, pp. 396–7.
54. This public lecture was originally published as 'The Influence of Women on the Progress of Knowledge', *Fraser's Magazine*, 57 (April 1858), pp. 395–407; it was later published in *Miscellaneous and Posthumous Works of Henry Thomas Buckle*, vol. 1, pp. 1–19. References will be given to the former work.
55. Buckle, 'The Influence of Women on the Progress of Knowledge', p. 396.
56. Ibid.
57. Ibid., p. 398.
58. Ibid., p. 400.
59. Ibid., p. 399.
60. Ibid., pp. 400–1.
61. Ibid., p. 400.
62. See, for instance, B. G. Smith, *Gender of History: Men, Women, and Historical Practice* (Cambridge, MA: Harvard University Press, 1998).
63. Buckle, *History of Civilization in England*, vol. 2, p. 396.
64. Whewell to Austin, 1 April 1858, in *Three Generations of English Women*, vol. 2, p. 56.
65. See, for instance, M. Ruse, 'Darwin's Debt to Philosophy: An Examination of the Influence of the Philosophical Ideas of John F. W. Herschel and William Whewell on the Development of Charles Darwin's Theory of Evolution', *Studies in the History and Philosophy of Science*, 6 (1975), pp. 159–81; Ruse, *The Darwinian Revolution: Science Red in Tooth and Claw* (1979), 2nd edn (Chicago, IL: University of Chicago Press, 1999), pp. 174–6, 179–80, 197–8; and D. L. Hull, *Darwin and His Critics: The Reception of Darwin's Theory of Evolution by the Scientific Community* (Cambridge, MA: Harvard University Press, 1973), ch. 2 ('The Inductive Method').
66. Darwin to Hooker, 23 February 1858, and Hooker to Darwin, 25 February 1858, in *The Correspondence of Charles Darwin*, vol. 7: *1858–1859*, ed. F. Burkhardt and S. Smith (Cambridge: Cambridge University Press, 1991), pp. 31, 34.

67. On the readings and reception of *Vestiges* see J. A. Secord, *Victorian Sensation: The Extraordinary Publication, Reception, and Secret Authorship of* Vestiges of the Natural History of Creation (Chicago, IL: University of Chicago Press, 2000).
68. On the comparison between the reception of *Vestiges* and of Buckle's *History of Civilization* see A. Desmond and J. Moore, *Darwin* (1991; London: Penguin Books, 1992), 436; but cf. H. Small, 'Chances Are', pp. 64–85 on p. 69 where it is argued that the anonymity of *Vestiges* was central to that book's reception whereas Buckle's persona was very much attached to the reception of his book.
69. C. Darwin, *The Life and Letters of Charles Darwin*, ed. F. Darwin, 2 vols (London: D. Appleton, 1887), vol. 2, p. 386.
70. G. J. Holyoake, *Sixty Years of an Agitator's Life*, 2 vols (London: T. Fisher Unwin, 1892), vol. 2, pp. 93–4.
71. Buckle, *History of Civilization*, vol. 2, p. 249.
72. Buckle to Emily Shirreff, 4 October 1858, Buckle Collection, MS 66, vol. 10, p. 163.
73. Buckle to Parker, 19 January 1856, Buckle Collection, MS 66, vol. 10, pp. 51–2; and Huth, *Life and Writings of Henry Thomas Buckle*, vol. 1, pp. 113–14.
74. Buckle to Parker, 22 February 1856, Buckle Collection, MS 66, vol. 10, p. 54; and Huth, *Life and Writings of Henry Thomas Buckle*, vol. 1, p. 114.
75. Buckle to Parker, 22 February 1856, Buckle Collection, MS 66, vol. 10, pp. 55–6; and Huth, *Life and Writings of Henry Thomas Buckle*, vol. 1, p. 116.
76. Buckle to Grey, 30 June 1856, Buckle Collection, MS 66, vol. 10, pp. 67–8.
77. Buckle to Mrs Grote, 17 May 1861, Buckle Collection, MS 66, vol. 1, p. 7; and MS 66, vol. 10, pp. 272–3.
78. Buckle, *History of Civilization in England*, vol. 2, p. 257.
79. Buckle to Mrs Huth, 16 April 1862, Buckle Collection, MS 66, vol. 2, p. 34; and MS 66 vol. 10, p. 342.
80. T. B. Sandwith to H. Huth, 14 December 1868, Buckle Collection, MS 66 vol. 3, p. 357.
81. On the plethora of criticisms directed at Buckle see J. M. Robertson, *Buckle and His Critics*; G. A. Wells, 'The Critics of Buckle', *Past and Present*, 9 (1956), pp. 75–89; and Parker, 'English Historians and the Opposition to Positivism'.
82. [G. Eliot], 'Rationalism in Europe', *Westminster Review*, 28:2 (October 1865), pp. 326–51. The book under review was W. E. H. Lecky, *History of the Rise and Influence of the Spirit of Rationalism in Europe*, 2 vols (London: Longmans, 1865). Lecky's later work abandoned Buckle's method in favour of a more inductive approach. See Semmel, 'H. T. Buckle', pp. 381–3; and D. McCartney, *W. E. H. Lecky: Historian and Politician, 1838–1903* (Dublin: The Lilliput Press, 1994), pp. 39–56. For Eliot's negative opinions of Buckle see N. McCaw, *George Eliot and Victorian Historiography: Imagining the National Past* (London: Macmillan Press, 2000), pp. 6–7.
83. [G. Eliot], Review of *The Progress of the Intellect, as Exemplified in the Religious Development of the Greeks and Hebrews* by R. W. Mackay, *Westminster Review*, 54:2 (January 1851), pp. 353–68.
84. [M. Pattison], 'Mackay's Tübingen School', *Westminster Review*, 24:2 (October 1863), pp. 510–31. On Mackay see G. Budge, 'Mackay, Robert William (1803–1882)', *Oxford Dictionary of National Biography* (Oxford: Oxford University Press, 2004). I greatly acknowledge the suggestion of an anonymous reviewer concerning the comparison between Mackay and Buckle.
85. Miss Rogers to H. Huth, 13 October 1866, Buckle Collection, MS 66, vol. 6, p. 13

86. [J. D. Lester], 'History and Biography', *Westminster Review*, 43 (1873), pp. 301–18 on p. 302.

2 The Sciences of History

1. William Stubbs to Edward A. Freeman, 8 November 1857, Bodleian Library, Oxford University (hereafter cited as Stubbs Letters), MS. Eng. misc. e. 148: 7.
2. On the opening of archives in Britain see P. Levine, *The Amateur and the Professional: Antiquarians, Historians and Archaeologists in Victorian England, 1838–1886* (Cambridge: Cambridge University Press, 1986).
3. Buckle, *History of Civilization in England*, vol. 2, p. 396.
4. William Stubbs to Edward A. Freeman, 13 April 1858, Stubbs Letters, MS. Eng. misc. e. 148: 9.
5. A. Pérez-Ramos, *Francis Bacon's Idea of Science and the Maker's Knowledge Tradition* (Oxford: Clarendon Press, 1988), p. 24. See also Yeo, 'An Idol of the Market-Place'; and Smith, *Fact and Feeling*, esp. ch. 1.
6. See, for instance, R. G. Collingwood, *The Idea of History* (Oxford: Oxford University Press, 1970), pp. 126–33; and G. S. Jones, 'History: The Poverty of Empiricism', in R. Blackburn (ed.), *Ideology in Social Science: Readings in Critical Theory* (London: Fontana/Collins, 1972), pp. 97–8.
7. Parker, 'English Historians and the Opposition to Positivism'.
8. M. Carignan, 'Analogical Reasoning in Victorian Historical Epistemology', *Journal of the History of Ideas*, 64 (2003), pp. 462–3.
9. T. W. Heyck, *The Transformation of Intellectual Life in Victorian England* (London/Canberra: St. Martin's Press, 1982), p. 137.
10. T. Gieryn, 'Boundary-Work and the Demarcation of Science from Non-Science: Strains and Interests in Professional Ideologies of Science', *American Sociological Review*, 48 (1983), p. 781. See also his *Cultural Boundaries of Science: Credibility on the Line* (Chicago, IL: University of Chicago Press, 1999).
11. R. Hill, *Lord Acton* (New Haven/New York: Yale University Press, 2000), ch. 2.
12. L. von Ranke, 'Preface: *Histories of the Latin and Germanic Nations from 1494–1514*', reprinted in F. Stern (ed.), *The Varieties of History from Voltaire to the Present* (New York: Vintage Books, 1973), p. 57.
13. For Ranke's views of impartiality see his 'On the Character of Historical Science (A Manuscript of the 1830s)', in Ranke, *The Theory and Practice of History*, ed. G. G. Iggers and K. von Moltke, trans. W. A. Iggers and Moltke (Indianapolis, NY: Bobbs-Merrill, 1973), pp. 41–3. See also G. G. Iggers, 'The Intellectual Foundations of Nineteenth-Century 'Scientific History': The German Model', in S. Macintyre, J. Maiguashca and A. Pók (eds), *The Oxford History of Historical Writing*, vol. 4: *1800–1945* (Oxford: Oxford University Press, forthcoming, 2011).
14. L. Krieger, *Ranke: The Meaning of History* (Chicago/London: The University of Chicago Press, 1977), p. 4.
15. Iggers, 'The Intellectual Foundations of Nineteenth-Century "Scientific History"'.
16. Lord Acton, 'The Study of History', in J. R. Fears (ed.), *Selected Writings of Lord Acton*, vol. 2: *Essays in the Study and Writing of History* (Indianapolis, IN: Liberty Fund, 1986), p. 533.
17. Ibid., p. 519.
18. Quoted in Hill, *Lord Acton*, p. 103.

19. Lord Acton, 'Döllinger's Historical Work', *English Historical Review*, 5 (1890), pp. 700–44 on p. 744.
20. See the essays in Benedikt Stuchtey and Peter Wende (eds), *British and German Historiography 1750–1950: Traditions, Perceptions, and Transfers* (Oxford/New York: Oxford University Press, 2000).
21. See, especially, Acton, 'The Catholic Press', *Rambler*, 11 (February 1859), pp. 73–90.
22. Hill, *Lord Acton*, p. 116.
23. Lord Acton to Richard Simpson, 30 March 1858, in *The Correspondence of Lord Acton and Richard Simpson*, ed. J. L. Altholz and D. McElrath, vol. 1 (Cambridge: Cambridge University Press, 1971), p. 21.
24. Acton to Simpson, 31 May 1858, ibid., p. 28.
25. Simpson to Acton, 9 June 1858, ibid., p. 32.
26. Lord Acton, 'Mr. Buckle's Thesis and Method (1858)', in *Essays in the Liberal Interpretation of History: Selected Papers*, ed. W. H. McNeill (Chicago/London: University of Chicago Press, 1967), p. 3.
27. Ibid., p. 6.
28. Ibid., p. 8.
29. Ibid., p. 11.
30. Ibid., pp. 8–9, 11.
31. Ibid., p. 11.
32. Ibid., p. 17.
33. Ibid., pp. 17–18.
34. Ibid., p. 18.
35. Ibid., p. 20.
36. Ibid., p. 27.
37. Ibid., p. 29.
38. Ibid., p. 30.
39. Ibid.
40. Ibid., p. 31.
41. Ibid., p. 40.
42. Ibid., pp. 23, 24.
43. On the German rejection of Buckle see Fuchs, *Henry Thomas Buckle*; and Fuchs, 'Contemporary Alternatives to German Historicism in the Nineteenth Century'.
44. J. G. Droysen, 'The Elevation of History to the Rank of Science', in *Outline of the Principles of History*, trans. E. B. Andrews (Boston, MA: Ginn and Company, 1893), pp. 61–89 on p. 66. The review was originally published as Droysen, 'Die Erhebung der Geschichte zum Rang einer Wissenschaft', *Historische Zeitschrift*, 9 (1862), 1–22.
45. Droysen, 'The Elevation of History to the Rank of Science', p. 76.
46. Lord Acton, 'Ultramontanism (1863)', in *Essays in the Liberal Interpretation of History*, p. 213. For the relationship between Acton's liberal Catholicism and his historical methodology see P. Hinchliff, 'Faith and History: Lord Acton and Catholic Modernism in Britain', in Hinchliff, *God and History: Aspects of British Theology 1875–1914* (Oxford: Clarendon Press, 1992), pp. 150–79, esp. pp. 156–60.
47. Acton to Simpson, 9 July 1858 and 16 July 1858, in *The Correspondence of Lord Acton and Richard Simpson*, pp. 52, 57.
48. Lord Acton, 'Mr. Goldwin Smith's *Irish History*', p. 69.
49. [G. Smith], Review of *History of England from the Fall of Wolsey to the Death of Elizabeth*, vols 1–4, by J. A. Froude, *Edinburgh Review*, 108 (July 1858), pp. 206–52 on p. 206.

50. Lord Acton to Mary Gladstone, 15 February 1884, in H. Paul (ed.), *Letters of Lord Acton to Mary Gladstone* (London: The Macmillan Company, 1904), p. 290.
51. The preceding biographical information is based on J. Campbell, 'Stubbs, William (1825–1901)', in *Oxford Dictionary of National Biography* (Oxford: Oxford University Press, 2004).
52. Stubbs to Freeman, 23 December [1865], Stubbs Letters, MS. Eng. misc. e. 148: 147.
53. Stubbs to Freeman, 4 August [1866], Stubbs Letters, MS. Eng. misc. e. 148: 160.
54. The following account is drawn from N. J. Williams, 'Stubbs's Appointment as Regius Professor, 1866', *Bulletin of the Institution of Historical Research*, 33 (1960), pp. 121–5.
55. H. L. Mansel to Lord Carnarvon, 15 July 1866, ibid., p. 123.
56. On the role of tutors and professors in the teaching of history at Oxford and Cambridge see R. N. Soffer, *Discipline and Power: The University, History, and the Making of an English Elite, 1870–1930* (Stanford, CA: Stanford University Press, 1994), esp. ch. 6.
57. On the *Saturday Review* see M. M. Bevington, *The Saturday Review, 1855–1868: Representative Educated Opinion In Victorian England* (New York: Columbia University Press, 1941), esp. ch. 8 (for the analysis of the *Saturday*'s treatment of historical subjects), pp. 342–46 (for articles attributed to Freeman) and pp. 349–50 (for articles attributed to Green).
58. [J. R. Green], 'Professor Stubbs's Inaugural Lecture', *Saturday Review*, 2 March 1867, p. 279.
59. Stubbs, 'Inaugural', p. 12.
60. Ibid., p. 13, emphasis added.
61. Ibid., p. 19.
62. Ibid., p. 1.
63. Ibid., p. 11.
64. Ibid., pp. 11–12.
65. Ibid., p. 12.
66. Stubbs to Freeman, 2 March 1864, Stubbs Letters, MS. Eng. misc. e. 148: 126.
67. J. R. Green to E. A. Freeman, 12 February 1867, in L. Stephen (ed.), *Letters of John Richard Green* (New York: The Macmillan Company, 1901), pp. 174–7.
68. Ibid., p. 23.
69. On natural theology see, in particular, J. Brooke, *Science and Religion: Some Historical Perspectives* (Cambridge: Cambridge University Press, 1991), esp. ch. 6.
70. Levine, *The Amateur and the Professional*, pp. 169–70; J. P. von Arx, *Progress and Pessimism*, pp. 198–9; and Hinchliff, 'Background: Historical and Religious Understanding in Nineteenth Century Britain', in Hinchliff, *God and History*, pp. 3–30, esp. pp. 13–17.
71. S. Bann, *The Inventions of History: Essays on the Representation of the Past* (Manchester/New York: Manchester University Press, 1990), p. 28.
72. W. Stubbs, 'The Comparative Constitutional History of Medieval Europe', in Stubbs, *Lectures on Early English History*, ed. A. Hassall (London: Longmans, Green, and Co., 1906), pp. 194–5.
73. Ibid., p. 195. See also M. Bentley, *Modernizing England's Past: English Historiography in the Age of Modernism, 1870–1970* (Cambridge: Cambridge University Press, 2005), 48.
74. W. Stubbs, *The Constitutional History of England in Its Origins and Development*, vol. 3 (1878) library edn (Oxford: Clarendon Press, 1880), p. 668.
75. This reading of Stubbs's *Constitutional History* is indebted to J. W. Burrow, *A Liberal Descent: Victorian Historians and the English Past* (Cambridge: Cambridge University Press, 1981), ch. 6, esp. pp. 144–9; and Bentley, *Modernizing England's Past*, 25. See also

Burrow, *A History of Histories: Epics, Chronicles, Romances and Inquiries from Herodotus and Thucydides to the Twentieth Century* (London: Penguin, 2007), pp. 405–12; and J. Vernon, 'Narrating the Constitution: The Discourse of "the Real" and the Fantasies of Nineteenth-Century Constitutional History', in Vernon (ed.), *Re-Reading the Constitution: New Narratives in the Political History of England's Long Nineteenth Century* (Cambridge: Cambridge University Press, 1996), pp. 204–29, esp. pp. 216–19, 223–6. On the developmental approach in Victorian history-writing see D. Goldstein, 'Confronting Time: The Oxford School of History and the Non-Darwinian Revolution', *Storia della Storiografia*, 45 (2004), pp. 3–27.
76. Stubbs, *The Constitutional History of England*, p. 2.
77. Ibid., p. 11.
78. Buckle, *History of Civilization in England*, vol. 1, pp. 29–30 and n. 1. The quotation that Buckle cited is from Mill's *Principles of Political Economy*, vol. 1 (London, 1849), p. 390.
79. Stubbs, *The Constitutional History of England*, vol. 1, p. 11.
80. M. Müller, *Lectures on the Science of Language* (London: Longman, Green, Longman & Roberts, 1861).
81. H. A. MacDougall, *Racial Myth in English History: Trojans, Teutons, and Anglo-Saxons* (Montreal, QC: Harvest House, 1982), ch. 7. See also P. B. Rich, *Race and Empire in British Politics* (1986) 2nd edn (Cambridge: Cambridge University Press, 1990), ch. 1.
82. E. A. Freeman, *The History of the Norman Conquest of England, its Causes and its Results*, vol. 1 (1867) 2nd edn (London: Macmillan, 1870), p. xii. For the pre-history of this debate in the seventeenth century see J. G. A. Pocock, *The Ancient Constitution and the Feudal Law: A Study of English Historical Thought in the Seventeenth Century* (Cambridge: Cambridge University Press, 1957).
83. Freeman, *The History of the Norman Conquest of England*, vol. 1, p. 1.
84. Burrow, *A Liberal Descent*, pp. 158–60.
85. Freeman, *The History of the Norman Conquest of England*, vol. 1, p. xiii.
86. Ibid., vol. 1, p. 2.
87. Ibid., vol. 1, p. 9. See also Burrow, *A Liberal Descent*, p. 128.
88. See in particular Freeman's *Comparative Politics* (London: Macmillan and Co., 1874); and S. Collini, D. Winch and J. Burrow, *That Noble Science of Politics: A Study in Nineteenth-Century Intellectual History* (Cambridge: Cambridge University Press, 1983), ch. 7.
89. [E. A. Freeman], Review of *Revolutions in English History* by Dr Vaughan, *Edinburgh Review*, 112 (July 1860), pp. 136–60 on pp. 158–60. This review was reprinted by Freeman under the title 'The Continuity of English History', in his *Historical Essays*, vol. 1 (1871) 5th edn (London: Macmillan, 1896), pp. 40–52 on pp. 50–2.
90. J. R. Seeley, 'History and Politics I', *Macmillan's Magazine*, 40 (August, 1879), p. 298.

3 Controversial Boys

1. Stubbs, *Seventeen Lectures on the Study of Medieval and Modern History*, p. 53; see also J. W. Burrow, *A Liberal Descent: Victorian Historians and the English Past* (Cambridge: Cambridge University Press, 1981), p. 97.
2. On the Tractarian or Oxford Movement see esp. O. Chadwick, *The Mind of the Oxford Movement* (Stanford, CA: Stanford University Press, 1961); and P. B. Nockles, *The Oxford Movement in Context: Anglican High Churchmanship, 1760–1857* (Cambridge: Cambridge University Press, 1994).

3. N. Vance, 'Kingsley, Charles (1819–1875)', in *Oxford Dictionary of National Biography* (Oxford: Oxford University Press, 2004).
4. O. Chadwick, 'Charles Kingsley at Cambridge', *The Historical Journal*, 18 (1975), pp. 303–25 on p. 307.
5. C. Kingsley, *Alton Locke, Tailor Poet* (1850) 3rd edn (London: Chapman and Hall, 1852), ch. 13.
6. It is important to note that most supposed members of the Broad Church would have refuted the label. See P. Corsi, *Science and Religion: Baden Powell and the Anglican Debate* (Cambridge: Cambridge University Press, 1988), pp. 198–9.
7. C. Darwin, *On the Origin of Species*, 6th edn (1872; New York: Mentor, 1958), p. 452; C. Kingsley to C. Darwin, in *Correspondence of Charles Darwin*, vol. 7: *1858–1859*, ed. Burkhardt and Smith, p. 380. See also J. Browne, *Charles Darwin*, vol. 2: *The Power of Place* (London: Pimlico, 2003), p. 95.
8. J. L. Altholz, *Anatomy of a Controversy: The Debate over Essays and Reviews, 1860–1864* (Aldershot: Scolar Press, 1994).
9. [T. C. Sanders], 'Two Years Ago', *Saturday Review*, 21 February 1857, p. 176–7 on p. 176.
10. 'Mr. Kingsley on the Science of History', *Literary Gazette*, 15 December 1860, pp. 509–10 on p. 509.
11. William Stubbs to Edward Freeman, 31 May 1860, Stubbs Letters, MS. Eng. misc. e. 148: 66.
12. [E. A. Freeman], 'Mr. Kingsley's Roman and Teuton', *Saturday Review*, 9 April 1864, pp. 446–8 on p. 446.
13. Ibid., p. 448.
14. C. Kingsley, 'The Limits of Exact Science as Applied to History', in *The Roman and the Teuton* (London: Macmillan and Co., 1864), p. ix.
15. Ibid., pp. xiii–xiv.
16. Ibid., pp. xiv–xv.
17. Ibid., p. xxxvii.
18. Ibid., p. xxxviii.
19. Ibid., p. xxxix.
20. Ibid., p. xl.
21. Ibid., p. xliv.
22. Ibid., p. xi.
23. Ibid., p. xii.
24. [H. Lawrance], Review of *The Limits of Exact Science as applied to History* by Charles Kingsley, *British Quarterly Review*, 33:65 (1861), pp. 264–7 on pp. 264–5.
25. 'Mr. Kingsley on the Science of History', p. 509.
26. Ibid., p. 510.
27. 'Mr. Kingsley on Science and History', *London Review*, 22 December 1860, pp. 99–600 on p. 599.
28. Ibid., p. 600.
29. [E. S. Beesly], 'Mr. Kingsley on the Study of History', *Westminster Review*, 75 (1861), pp. 305–36 on pp. 309 (history and biography), 312 (science and history).
30. Ibid., pp. 307, 312.
31. [Freeman], 'Mr. Kingsley's Roman and Teuton', p. 446.

32. Ibid., p. 447; see also, M. M. Bevington, *The Saturday Review 1855–1868: Representative Educated Opinion in Victorian England* (New York: Columbia University Press, 1941), ch. 8.
33. [E. P. Hood], 'The Roman and the Teuton', *Eclectic Review*, 7 (July 1864), pp. 82–8 on p. 82.
34. Ibid., p. 83.
35. Kingsley, *The Roman and the Teuton*, p. 1.
36. F. M. Müller, 'Preface', in Kingsley, *The Roman and the Teuton* 2nd edn (London: Macmillan and Company, 1891), pp. x–xi.
37. Ibid., pp. xii–xiii.
38. On the popularity of Kingsley's lectures see G. Kitson Clark, 'A Hundred Years of the Teaching of History as Cambridge, 1873–1973', *The Historical Journal*, 16 (1973), p. 535; J. O. McLachlan, 'The Origin and Early Development of the Cambridge Historical Tripos', *Cambridge Historical Journal*, 9 (1947), pp. 79–80; and Chadwick, 'Charles Kingsley at Cambridge', p. 311.
39. J. A. Froude, *The History of England from the Death of Cardinal Wolsey to the Defeat of the Spanish Armada*, 12 vols (London: Longmans, Green, and Co., 1856–70), vol. 1, p. 91. It should be noted that Froude initially entitled the work *The History of England from the Fall of Wolsey to the Death of Elizabeth* but decided later to end the narrative at the height of Elizabeth's power with the defeat of the Spanish Armada.
40. R. H. Froude, *Remains of the Late Reverend Richard Hurrell Froude*, ed. J. H. Newman and J. Keble, 4 vols (London: J. G. & F. Rivington, 1838); and see P. Brendon, *Hurrell Froude and the Oxford Movement* (London: Paul Elek, 1974).
41. Ibid., p. 183.
42. J. Markus, *J. Anthony Froude: The Last Undiscovered Great Victorian* (New York: Scribner, 2005), p. 15.
43. Quoted in Brendon, *Hurrell Froude and the Oxford Movement*, p. 180.
44. J. Markus, *J. Anthony Froude*, p. 15. On the reaction of the public to Froude's *Remains* see Brendon, *Hurrell Froude and the Oxford Movement*, ch. 15; Nockles, *The Oxford Movement in Context*, pp. 281–2; and O. Chadwick, *The Victorian Church*, vol. 1 (London: Adam & Charles Black, 1966), pp. 172–81.
45. J. A. Froude, 'A Few Words on Mr. Freeman', *Nineteenth Century*, 5 (1879), pp. 618–37 on p. 621.
46. Quoted in Markus, *J. Anthony Froude*, p. 37.
47. W. H. Dunn, *James Anthony Froude: A Biography*, 2 vols (Oxford: Clardendon Press, 1963), vol. 1, p. 131; and J. W. Burrow, *A Liberal Descent: Victorian Historians and the English Past* (Cambridge: Cambridge University Press, 1981), p. 235.
48. J. A. Froude, *The Nemesis of Faith* (London: John Chapman, 1849), p. 11.
49. Ibid., p. 17.
50. Ibid., p. 22.
51. Ibid., p. 39.
52. Ibid., p. 6.
53. Dunn, *James Anthony Froude*, vol. 1, p. 126; and Burrow, *A Liberal Descent*, p. 257.
54. Froude, *The Nemesis of Faith*, p. 43.
55. [J. A. Froude], 'England's Forgotten Worthies', *Westminster Review*, 58 (1852), pp. 32–67; and see the discussion by Markus, *J. Anthony Froude*, pp. 62–3.
56. [Smith], Review of *History of England*, p. 206.

57. On the opening of archives in Britain see P. Levine, *The Amateur and the Professional: Antiquarians, Historians and Archaeologists in Victorian England, 1838–1886* (Cambridge: Cambridge University Press, 1986).
58. [Smith], Review of *History of England*, p. 208.
59. Ibid., pp. 211–12.
60. Ibid., p. 212.
61. Ibid., p. 252.
62. J. A. Froude, 'The "Edinburgh Review" and Mr. Froude's History', *Frazer's Magazine*, 58 (September 1858), p. 363.
63. It would not help matters that Froude would later write a controversial biography of Carlyle confirming his devotion to the doctrine of hero worship. J. A. Froude, *Thomas Carlyle*, 2 vols (London: Longmans, Green, 1882, 1884); and W. H. Dunn, *Froude and Carlyle: A Study of the Froude–Carlyle Controversy* (1930; Port Washington, NY: Kennikat Press, 1969). On Carlyle's influence on Froude see Burrow, *A Liberal Descent*, pp. 251–7.
64. Froude, 'The Science of History', pp. 1, 2.
65. Ibid., p. 14.
66. Ibid., p. 19.
67. Ibid., p. 20.
68. Ibid., p. 21.
69. Ibid., pp. 36–7.
70. J. Morley, 'Mr. Froude on the Science of History', *Fortnightly Review*, 2 (1867), pp. 226–37; 'Mr. Froude on the Science of History', *The Reader*, 27 February 1864, pp. 255–6; and 'Mr. Froude's Short Studies on Great Subjects', *The London Review*, 14:353 (6 April 1867), pp. 406–7.
71. 'Mr. Froude's Short Studies on Great Subjects', p. 406.

4 Discipline and Disease

1. Kingsley told J. R. Green that he gave up the Chair because Freeman's 'criticisms had made him feel he never ought to have taken it'. J. R. Green to E. A. Freeman, 11 November 1875, in L. Stephen (ed.), *Letters of John Richard Green* (New York: The Macmillan Company, 1901), p. 423. For Freeman's criticisms of Kingsley see [E. A. Freeman], 'Mr. Kingsley's Roman and Teuton', *Saturday Review*, 9 April 1864, pp. 446–8; [Freeman], 'Kingsley's Hereward', *Saturday Review*, 19 May 1866, pp. 594–5; and [Freeman], 'Professor Kingsley on the "Ancien Regime"', *Saturday Review*, 22 June 1867, pp. 792–3. See also O. Chadwick, 'Charles Kingsley at Cambridge', *Historical Journal*, 18 (1975), pp. 303–25; L. Stephen, *Some Early Impressions* (London: Leonard & Virginia Woolf at the Hogarth Press, 1924), p. 42; and N. Vance, 'Kingsley, Charles (1819–1875)', in *Oxford Dictionary of National Biography* (Oxford: Oxford University Press, 2004).
2. On the relationship between science and Christianity in Seeley's thought see S. Rothblatt, *The Revolution of the Dons: Cambridge and Society in Victorian England* (London: Faber and Faber, 1968), pp. 162–5, 172–5.
3. J. R. Seeley, *Ecce Homo: A Survey of the Life and Work of Jesus Christ* (1865; London: Macmillan, 1912), pp. 367–8.
4. See, for instance, 'Ecce Homo', *The Spectator*, 23 December 1865, p. 1436.
5. See, for instance, [W. Elwin], 'Ecce Homo', *Quarterly Review*, 119 (1866), pp. 515–29.

6. D. Wormell, *Sir John Robert Seeley and the Uses of History* (Cambridge: Cambridge University Press, 1980), ch.1.
7. See, for instance, the letters from A. Shawn, 20 November 1882 and W. H. Johnston, 3 June 1866, J. R. Seeley Papers, University of London, UK (hereafter cited as Seeley Papers), MS 903/3A/4.
8. J. Christien to the author of *Ecce Homo*, 22 May 66, Seeley Papers, MS 903/3A/4.
9. Kingsley's wife is quoted in Macmillan to Seeley, 29 March 1866, Seeley Papers, MS 903/3A/1; and Rose Kingsley to Macmillan, 7 March 1866, Seeley Papers, MS 903/3A/2.
10. For reference to this letter see Macmillan to Gladstone, 29 December 1865, Seeley Papers, MS 903/3A/1; and W. E. Gladstone, *On Ecce Homo* (London: Strahan & Co. Publishers, 1868).
11. Gladstone, *On Ecce Homo*, p. 1.
12. I. Hesketh, 'The Victorian Bible: *Ecce Homo* and the Manufacturing of a Literary Sensation', *Canadian Society of Church History Annual Conference*, June 2008, Vancouver, BC.
13. 'Literary Gossip', *The London Review*, 17 November 1866, pp. 559–60 on p. 560.
14. Macmillan to Seeley, 12 December 1866, Seeley Papers, MS 903/3A/1.
15. Seeley to Macmillan, 7 July 1872, Macmillan Archive, British Library, London UK (hereafter cited as Macmillan Archive), Add MS 55074: 8.
16. 'Modern History at Cambridge', *Saturday Review*, 19 February 1870, pp. 241–2 on p. 241. On Gladstone's appointment of Seeley see Rothblatt, *The Revolution of the Dons*, pp. 176–8.
17. 'Modern History at Cambridge', p. 241.
18. 'Our Cambridge Letter', *Athenaeum*, 26 February 1870, p. 291.
19. 'Modern History at Cambridge', p. 242.
20. Ibid., p. 241.
21. J. R. Green to E. A. Freeman, 5 March 1870, in Stephen (ed.), *Letters of John Richard Green*, p. 240.
22. G. R. Elton, *Return to Essentials: Some Reflections on the Present State of Historical Study* (Cambridge: Cambridge University Press, 2002), p. 102; and Butler and Elton quoted in Vance, 'Kingsley, Charles (1819–1875)'.
23. 'Seeley's Lectures and Essays', *Saturday Review*, 7 January 1871, pp. 23–4 on p. 23.
24. See R. N. Soffer, *Discipline and Power: The University, History, and the Making of an English Elite, 1870–1930* (Stanford, CA: Stanford University Press, 1994), p. 56.
25. E. A. Freeman, 'Inaugural Lecture: The Office of the Historical Professor', in *The Methods of Historical Study* (London: The Macmillan Company, 1886), p. 22.
26. 'Seeley's Life and Times of Stein', *Saturday Review*, 1 February 1879, pp. 146–7.
27. J. R. Seeley, 'History and Politics I', *Macmillan's Magazine*, 40 (August 1879), p. 289.
28. Ibid., pp. 296–9 on p. 298.
29. Ibid., p. 294.
30. Ibid., pp. 289–90.
31. Ibid., p. 290.
32. Ibid., p. 290.
33. Ibid., pp. 291–2.
34. Ibid., p. 291.
35. Ibid., p. 292.
36. Ibid., p. 293.
37. Ibid., p. 294.

38. J. R. Seeley, 'History and Politics IV', *Macmillan's Magazine*, 41 (1879), p. 31.
39. E. A. Freeman, 'On the Study of History', *Fortnightly Review*, 35 (1881), pp. 319–39 on p. 326.
40. Ibid., p. 325.
41. E. A. Freeman, 'Lecture III: The Nature of Historical Evidence', in *The Methods of Historical Study* (London: The Macmillan Company, 1886), p. 118. See also R. Jann, *The Art and Science of Victorian History* (Columbus, OH: Ohio State University Press, 1985), p. 176.
42. Ibid., pp. 121–2.
43. Freeman, 'Lecture II: The Difficulties of Historical Study', in *The Methods of Historical Study*, pp. 99–100.
44. Freeman, 'The Difficulties of Historical Study', p. 103.
45. E. A. Freeman, 'The Mythical and Romantic Elements in Early English History', *Fortnightly Review*, 24 (1 May 1866), p. 644.
46. See, for instance, Review of *History of England* by J. R. Gardiner, *Athenaeum*, 21 March 1863, pp. 392–3. 'We do not every day meet an author with whom we could so easily agree, if, in reading him, we could only keep awake. Mr Gardiner's pulse is slow, and his paragraphs long.' Ibid., p. 392. Cf. [B. M. Cordery], 'Gardiner's *Reign of James I*', *Edinburgh Review*, 143: 291 (January 1876), pp. 101–41.
47. Freeman to Miss Edith Thompson, 15 January 1889, in W. R. W. Stephens, *Life and Letters of Edward A. Freeman*, 2 vols (London: Macmillan Company, 1895), vol. 2, p. 393.
48. Freeman to Miss Edith Thompson, 17 December 1882, ibid., vol. 2, p. 266.
49. See Freeman to J. Bryce, 316 September 1866, Papers of James Bryce (hereafter cited as Bryce Papers), MSS 5, X-films 8/31: 116.
50. D. M. Owen, 'The Chichele Professorship of Modern History, 1862', *Bulletin of the Institute of Historical Research*, 34 (1961), pp. 217–20.
51. Ibid.
52. 'E. A. Freeman to the electors for the Chichele Professorship of Modern History', reprinted ibid., p. 218–19.
53. P. G. Medd quoted ibid., p. 217.
54. 'The Oxford Professoriate', *Saturday Review*, 12 July 1862, p. 46.
55. Ibid., p. 46.
56. M. Burrows, *Autobiography of Montagu Burrows*, ed. S. Montagu Burrows (London: Macmillan and Company, 1908), p. 216. See also R. N. Soffer, *Discipline and Power: The University, History, and the Making of an English Elite, 1870–1930* (Stanford, CA: Stanford University Press, 1994), pp. 84–5.
57. T. F. Gieryn, 'Boundary-Work and the Demarcation of Science from Non-Science: Strains and Interests in Professional Ideologies of Scientists', *American Sociological Review*, 48 (1983), pp. 781–95 on pp. 791–2.
58. Ibid., pp. 787–9. See also Gieryn, *Cultural Boundaries of Science* (Chicago, IL: University of Chicago press, 1999), ch. 1, for the way in which John Tyndall sought to demarcate science from religion on the one hand and practical mechanics on the other. The issue of the relationship between science and its publics and the construction and maintenance of the boundaries separating scientific communities in Victorian Britain has virtually developed into a field of its own. See, in particular, R. Cooter, *The Cultural Meaning of Popular Science: Phrenology and the Organization of Consent in Nineteenth-Century Britain* (Cambridge/New York: Cambridge University Press, 1984); Cooter and S. Pumfrey, 'Separate Spheres and Public Places: Reflections on the History of Sci-

ence Popularization and Science in Public Culture', *History of Science*, 32 (1994), pp. 237–67; A. Winter, 'The Construction of Orthodoxies and Heterodoxies in the Early Victorian Life Sciences', in B. Lightman (ed.), *Victorian Science in Context* (Chicago, IL: University of Chicago Press, 1997), pp. 24–50; Winter, *Mesmerized: Powers of Mind in Victorian Britain* (Chicago, IL: University of Chicago Press, 1998); J. Second, *Victorian Sensation: The Extraordinary Publication, Reception, and Secret Authorship of* Vestiges of the Natural History of Creation (Chicago, IL: University of Chicago Press, 2000); A. Desmond, 'Redefining the X Axis: "Professionals", "Amateurs" and the Making of Mid-Victorian Biology – A Progress Report', *Journal of the History of Biology*, 34 (2001), pp. 3–50; and B. Lightman, *Victorian Popularizers of Science: Designing Nature for New Audiences* (Chicago, IL: University of Chicago Press, 2007).
59. [E. A. Freeman], 'Froude's Reign of Elizabeth (First Notice)', *Saturday Review*, 16 January 1864, pp. 80–2 on p. 81.
60. Ibid.
61. E. A. Freeman, 'Mr. Froude's Life and Times of Thomas Beckett', *The Contemporary Review*, 31–3 (1878), vol. 31, pp. 821–42; vol. 32, pp. 116–39, 474–500; vol. 33, pp. 213–41. Froude's 'Life and Times of Thomas Becket' first appeared in the *Nineteenth Century* (1878) and then in Froude, *Short Studies on Great Subjects*, vol. 4, pp. 1–230.
62. Freeman, 'Mr. Froude's Life and Times of Thomas Beckett', vol. 31, p. 821.
63. Ibid., vol. 31, p. 825.
64. Ibid., vol. 33, p. 213.
65. Ibid., vol. 31, p. 826.
66. Ibid., vol. 31, p. 827.
67. Ibid., vol. 33, p. 241.
68. Ibid.
69. R. H. Froude, 'History of the Contest between Thomas à Becket, Archbishop of Canterbury, and Henry II., King of England', in *Remains of Richard Hurrell Froude*, ed. John Henry Newman and John Keble, vol. 2 (Derby: J. G. & F. Rivington, 1839).
70. Ibid., vol. 31, p. 822. On the theological issues underpinning Freeman's attacks see I. Hesketh, 'The Remains of the Freeman–Froude Controversy: The Religious Dimension', *Historical Papers 2010: Canadian Society of Church History* (forthcoming, 2010); R. McNeill, 'Froude and Freeman', *Monthly Review*, 22 (1906), pp. 79–91; and H. Paul, *The Life of Froude* (London: Sir Isaac Pitman & Sons, 1905), ch. 5.
71. Froude, 'A Few Words on Mr. Freeman', p. 619.
72. Ibid., p. 620.
73. Ibid., p. 621.
74. L. Strachey, 'Froude', in *Portraits in Miniature and Other Essays* (London: Chatto & Windus, 1931), pp. 195–206 on p. 201. This essay was originally published as Strachey, 'One of the Victorians', *Saturday Review of Literature*, 6 December 1930, pp. 418–19.
75. J. Markus, *J. Anthony Froude: The Last Undiscovered Great Victorian* (New York: Scribner, 2005), p. 181.
76. Froude, 'A Few Words on Mr. Freeman', pp. 620, 627.
77. Markus, *J. Anthony Froude*, p. 182.
78. J. Bryce to E. A. Freeman, 3 April 1879, Bryce Papers, MSS 9, X-films 8/3.
79. W. Stubbs to E. A. Freeman, 2 April 1879, Stubbs Letters, MS. Eng. misc. e. 148: 258.
80. 'Some Papers in the Magazines', *The Spectator*, 5 April 1879, pp. 442–4 on p. 443.
81. E. A. Freeman, 'Last Words on Mr. Froude', *Contemporary Review*, 35 (1879), pp. 215–36 on p. 217.

82. Ibid., pp. 218–19.
83. Ibid., p. 219.
84. E. A. Freeman to James Bryce, 4 May 1979, Bryce Papers, MSS 6, X-films 8/5: 200.
85. E. A. Freeman to J. R. Green, 27 April 1879, E. A. Freeman Archive, John Rylands Library of Manchester, UK (hereafter cited as Freeman Archive), FA1/8/85.
86. Stubbs to Freeman, 18 February [1866], Stubbs Letters, MS. Eng. misc. e. 148: 258.
87. Stubbs to Freeman, quoted in Soffer, *Discipline and Power*, p. 86.
88. Stubbs, *Seventeen Lectures on the Study of Medieval and Modern History*, pp. v–vi.
89. Stubbs, 'A Last Statutory Public Lecture (8 May 1884)' ibid., pp. 372, 382.
90. M. Creighton addressed what he believed to be the inherent contradiction of professional historians as lecturers in 'The Endowment of Research', *Macmillan's Magazine*, 34 (1876), pp. 186–92.
91. E. A. Freeman to Goldwin Smith, 1884, in Stephens, *The Life and Letters of Edward A. Freeman*, vol. 2, p. 278.
92. Ibid., vol. 2, p. 291.
93. C. V. Langlois and C. Seignobos, *Introduction to the Study of History*, trans. G. G. Berry (New York: Henry Holt and Company, 1898), pp. 125–8.
94. F. York Powell, 'To the Reader', ibid., p. vi.
95. W. E. Spahr and R. J. Swenson, *Methods and Status of Scientific Research with Particular Applications to the Social Sciences* (New York: Harper & Brothers Publishing, 1930), p. 49. See also H. M. Stephens, *Counsel upon the Reading of Books* (1900; Port Washington, NY: Kennikat Press, 1968), pp. 81–2; F. L. Whitney, *The Elements of Research* (New York: Prentice-Hall, 1942), p. 199; and K. A. Nilakanta Sastri and H. S. Ramanna, *Historical Method in Relation to Indian History* (Madras: S. Viswanathan, 1956), p. 16.

5 History from Nowhere

1. On the ideals of manliness in nineteenth-century Britain see, for instance, J. Tosh, *Manliness and Masculinities in Nineteenth-Century Britain* (Harlow: Pearson, 2005), pp. 83–102.
2. 'Lord Acton's Inaugural Lecture', *Speaker*, 15 June 1895, pp. 648–9 on p. 648.
3. 'Lord Acton on Trial', *Saturday Review*, 22 June 1895, pp. 821–2 on p. 821.
4. W. L. Lilly, 'The New Spirit in History', *Nineteenth Century*, 38 (1895), pp. 619–33 on p. 619.
5. Ibid., p. 620.
6. The *Saturday Review* seems to be the exception. It was disappointed by Acton's confusing lecture and his cosmopolitan style of speech. 'Lord Acton on Trial', *Saturday Review*, 22 June 1895, pp. 821–2.
7. On this shift at Cambridge see in particular S. Rothblatt, *The Revolution of the Dons: Cambridge and Society in Victorian England* (London: Faber and Faber, 1968).
8. Lord Acton, 'The Study of History [1895]', in *Selected Writings of Lord Acton*, vol. 2: *Essays in the Study and Writing of History*, ed. J. R. Fears (Indianapolis, IN: Liberty Classics, 1986), pp. 504–52 on p. 504.
9. Ibid., p. 519.
10. Ibid., p. 533.
11. Ibid., p. 520.
12. Ibid., p. 552.

13. L. Daston and P. Galison, *Objectivity* (New York: Zone Books, 2007). See also Daston and Galison, 'The Image of Objectivity', *Representations*, 40 (1992), pp. 81–128; and Daston, 'Objectivity and the Escape from Perspective', *Social Studies of Science*, 22 (1992), pp. 597–619.
14. T. M. Porter, 'The Objective Self', *Victorian Studies*, 50 (2008), pp. 641–7 on p. 645. See also Porter's analysis of Pearson's attempts to construct an objective scientific self in *Karl Pearson: The Scientific Life in a Statistical Age* (Princeton, NJ: Princeton University Press, 2004); and on Huxley's strategies of self-presentation see P. White, *Thomas Huxley: Making the 'Man of Science'* (Cambridge: Cambridge University Press, 2003).
15. Acton, 'The Study of History [1895]', pp. 545–6.
16. Ibid., p. 552, emphasis added.
17. T. Nagel, *The View from Nowhere* (Oxford: Oxford University Press, 1986).
18. G. Levine, *Dying to Know: Scientific Epistemology and Narrative in Victorian England* (Chicago, IL: University of Chicago Press, 2002).
19. Acton, 'The Study of History [1895]', p. 534.
20. 'Lord Acton's Inaugural Lecture', *Speaker*, p. 649.
21. On the origins of the *English Historical Review* see D. S. Goldstein, 'The Origins and Early Years of the *English Historical Review*', *English Historical Review*, 101 (1986), pp. 6–19; A. Kadish, 'Scholarly Exclusiveness and the Foundation of the *English Historical Review*', *Historical Research*, 61 (1988), pp. 183–98; and L. Howsam, 'Academic Discipline or Literary Genre? The Establishment of Boundaries in Historical Writing', *Victorian Literature and Culture* (2004), pp. 525–45 on pp. 29–34.
22. L. Stephen (ed.), *Letters of John Richard Green* (New York: Macmillan and Company, 1901), p. 173; and Kadish, 'Scholarly Exclusiveness and the Foundation of the *English Historical Review*', p. 189.
23. J. R. Green to A. Macmillan, 4 February 1870, Macmillan Archive, Add MS 55058: 12.
24. C. M. D. Crowder, 'Creighton, Mandell (1843–1901)', *Oxford Dictionary of National Biography* (Oxford: Oxford University Press, 2004).
25. On Creighton and Acton's correspondence see F. Engel de Janösi, 'The Correspondence between Lord Acton and Bishop Creighton', *Cambridge Historical Journal*, 6 (1939), pp. 306–21.
26. Mandell Creighton to Lord Acton, 9 December 1882, Acton Papers, University Library, Cambridge University, UK (hereafter cited as Acton Papers), Add MS 6871: 1.
27. Lord Acton, Review of *A History of the Papacy during the Period of the Reformation* by Mandell Creighton, *Academy*, 9 December 1882, pp. 407–9 on pp. 408, 407.
28. Ibid., p. 407.
29. Creighton to Lord Acton, 9 December 1882, Acton Papers, Add MS 6871: 1
30. Lord Acton to Mandell Creighton, 14 December 1882, Acton Papers, Add MS 6871: 3.
31. D. R. Kelley, *Fortunes of History: Historical Inquiry from Herder to Huizinga* (New Haven, CT: Yale University Press, 2003), p. 232.
32. Mandell Creighton to Lord Acton, 17 July 1885 Acton Papers, Add MS 8119(1): C248. Creighton did manage to get Freeman to write something on 'a subject as yet untouched' (preferably something on the fifth century) in hopes of 'attract[ing] more attention'. See Creighton to Freeman, 3 August 1885, Freeman Archive, MS 1/7/123a; and E. A. Freeman, 'The Tyrants of Britain, Gaul, and Spain, A.D. 406–411', *English Historical Review*, 1 (1886), pp. 53–85.
33. Acton to Creighton, 14 August 1885, Acton Papers, Add MS 6871: 13.

34. Acton to Creighton, 9 September 1885, Acton Papers, Add MS 6871: 17.
35. Creighton to Acton, 14 September 1885, Acton Papers, Add MS 8119 (1). It seems that Creighton may have initially approached Acton about writing an introductory statement even earlier. He wrote to Bryce earlier in the summer that he approached Acton about such an introduction. Mandel Creighton to James Bryce, 31 July 1885, Bryce Papers, MS 53, X-films 8/5: 145.
36. Acton to Creighton, 17 September 1885, Acton Papers, Add MS 6871: 19.
37. The article was so late in fact that Creighton requested that Acton send it directly to the printer. See Creighton to Bryce, 10 December 1885, Papers of James Bryce, MS 53, X-films 8/5: 149.
38. Creighton to Acton, 23 December 1885, Acton Papers, Add MS 8119(1): C250.
39. Ibid.; and Acton to Creighton, December 1885, Acton Papers, Add MS 6871: 27.
40. Creighton to Bryce, 31 October 1885, Bryce Papers, MS 53, X-films 8/5: 147.
41. Creighton to Bryce, 14 December 1885, Bryce Papers, MS 53, X-films 8/5: 149.
42. C. Harvie, 'Bryce, James, Viscount Bryce (1838–1922)', *Oxford Dictionary of National Biography* (Oxford: Oxford University Press, 2004).
43. Creighton to Bryce, 20 December 1885, Bryce Papers, MS 53, X-films 8/5: 151.
44. Creighton to Bryce, 28 December 1885, Bryce Papers, MS 53, X-films 8/5: 153.
45. [James Bryce], 'Prefatory Note', *English Historical Review*, 1 (1886), pp. 1–6 on 1.
46. Ibid., p. 2.
47. Ibid.
48. Ibid., p. 4.
49. Ibid., p. 5.
50. On the exclusion of Froude from the review see Kadish, 'Scholarly Exclusiveness and the Foundation of the *English Historical Review*', pp. 190–1, 194–5 and 197–8.
51. Acton to Creighton, 16 November 1885, Acton Papers, Add MS 6871: 23.
52. Lord Acton, 'German Schools of History', *English Historical Review*, 1 (1886), pp. 7–42 on p. 42.
53. Ibid., p. 13–14.
54. Creighton to Acton, 26 January 1886, Acton Papers, Add MS 8119(1): C250.
55. Creighton to Bryce, 28 December 1885, Bryce Papers, MS 53, X-films 8/5:153.
56. 'The English Historical Review', *Saturday Review*, 23 January 1886, p. 128.
57. Acton to Creighton, 19 January 1886, Acton Papers, Add MS 6871.
58. Creighton to Acton, 24 January 1887, Acton Papers, Add MS 8119(1): C255.
59. Acton to Creighton, 11 March 1887, Acton Papers, Add MS 6871: 45.
60. Acton to Creighton, 5 April 1887, Acton Papers, Add MS 6871: 53.
61. Creighton to Acton, 9 April 1887, Acton Papers, Add MS 6871: 65.
62. Creighton to R. L. Poole, 5 April 1887, in L. Creighton, *Life and Letters of Mandell Creighton*, 2 vols (London: Longmans, Green and Company, 1904), vol. 1, p. 370.
63. Creighton to Freeman, 1 August 1887, Freeman Archive, 1/7/124.
64. See the discussion in Creighton to Acton, 7 June 1887, 14 June 1887 and 17 June 1887, Acton Papers, Add MS 8119(1): C257–9.
65. Lord Acton, Review of *A History of the Papacy*, p. 578.
66. On Acton and the *Cambridge Modern History* see L. Howsam, *Past into Print: The Publishing of History in Britain 1850–1950* (London: The British Library; Toronto: University of Toronto Press, 2009), pp. 62–71; G. N. Clark, 'The Origin of the *Cambridge Modern History*', *The Cambridge Historical Journal*, 8 (1945), pp. 57–64; and J.

L. Altholz, 'Lord Acton and the Plan of the *Cambridge Modern History*', *The Historical Journal*, 39 (1996), pp. 723–36.
67. Acton's report on the *Cambridge Modern History*, 1897, Acton Papers, Add MS 7729: F(1).
68. Lord Acton, 'Letter to Contributors to the *Cambridge Modern History*', in *Essays in the Liberal Interpretation of History*, ed. W. H. McNeill (Chicago, IL: University of Chicago Press, 1967), pp. 396–9 on p. 397.
69. Ibid., 399. See also Lord Acton, 'The *Cambridge Modern History*', in *Selected Writings of Lord Acton*, vol. 2, pp. 675–86 for the original proposal to the Cambridge Press syndics.
70. Roland Barthes, 'The Discourse of the Real', trans. S. Bann, in E. S. Shaffer (ed.), *Comparative Criticism: A Yearbook*, vol. 3 (Cambridge: Cambridge University Press, 1981), pp. 3–20 on pp. 11, 17. For an application of this theory to Victorian history-writing see J. Vernon, 'Narrative the Constitution: The Discourse of "the Real" and the Fantasies of Nineteenth-Century Constitutional History' in Vernon (ed.), *Re-Reading the Constitution: New Narratives in the Political History of England's Long Nineteenth Century* (Cambridge: Cambridge University Press, 1996), pp. 204–38.
71. Daston makes the same claim about scientists in 'Objectivity and the Escape from Perspective', p. 614.
72. J. A. Secord, *Victorian Sensation: The Extraordinary Publication, Reception, and Secret Authorship of* Vestiges of the Natural History of Creation (Chicago, IL: University of Chicago Press, 2000), pp. 19–20.
73. Acton to Creighton, 16 November 1896, Acton Papers, Add MS 6871: 102.
74. Acton to Creighton, 15 May 1898, Acton Papers, Add MS 6871: 112.
75. M. Creighton, 'Introductory Note', in A. W. Ward, G. W. Prothero and S. Leathes (eds), *The Cambridge Modern History*, vol. 1: *The Renaissance*, planned by Lord Acton (1902; Cambridge: Cambridge University Press, 1907), pp. 1–6 on p. 5.
76. Ibid., p. 1.
77. Ibid., p. 3.
78. Ibid., p. 4.
79. Ibid., p. 5.
80. Ibid., p. 6, 5.
81. W. Ward, G. W. Prothero, and S. Leathes, 'Preface', in *The Cambridge Modern History*, pp. v–viii on pp. v, vi.

6 Broad Shadows and Little Histories

1. S. Collini, *Public Moralists: Political Thought and Intellectual Life in Britain 1850–1930* (Oxford: Clarendon Press, 1991), ch. 6.
2. See, in particular, L. Howsam, 'Academic Discipline or Literary Genre? The Establishment of Boundaries in Historical Writing', *Victorian Literature and Culture* (2004), pp. 525–45.
3. 'Memorandum respecting the publication of Mr. Freeman's History of the Federal Government', Macmillan Archive, Add MS 55049: 1.
4. E. A. Freeman to A. Macmillan, 16 October 1863, Macmillan Archive, Add MS 55049: 6. Reference to the precise amount paid is in Freeman to Macmillan, 7 January 1864 and 22 November 1865, Macmillan Archive, Add MS 55049: 11, 35.
5. See Freeman to Macmillan, 9 November 1863 and 8 February 1864, Macmillan Archive, Add MS 55049: 9, 14.

6. Freeman to Macmillan, 26 November 1865 and 1 December 1865, Macmillan Archive, Add MS 55049: 37, 39.
7. Freeman to Macmillan, 16 October 1863, Macmillan Archive, Add MS 55049: 6.
8. Freeman to Macmillan, 14 November 1864, Macmillan Archive, Add MS 55049: 12.
9. Freeman to Macmillan, 1 December 1865, Macmillan Archive, Add MS 55049: 39.
10. Freeman to Macmillan, 4 December 1865, Macmillan Archive, Add MS 55049: 42.
11. [E. A. Freeman], 'Mr. Kingsley's Roman and Teuton', *Saturday Review*, 9 April 1864, pp. 446–8.
12. Freeman to Macmillan, 4 December 1865 and 14 January 1866, Macmillan Archive, Add MS 55049: 42, 44.
13. On Macmillan, Freeman and the Clarendon Press see L. Howsam, *Past into Print: The Publishing of History in Britain 1850–1950* (London: The British Library; Toronto: University of Toronto Press, 2009), 40–4.
14. Freeman to Macmillan, 13 May 1866, Macmillan Archive, Add MS 55049: 47.
15. 'Publishing Agreement for Old English History for Children', Macmillan Archive, Add MS 55049: 150
16. E. A. Freeman, *Old English History for Children* (London: Macmillan, 1869), p. v.
17. J. R. Green to E. A. Freeman, 1869, in Leslie Stephen (ed.), *Letters of John Richard Green* (New York: Macmillan and Company, 1901), p. 237.
18. Freeman to Macmillan, 15 April 1870, Macmillan Archive, Add MS 55049: 135.
19. Miss A. V. Ponsonby to Freeman, 18 October 1872, in W. R. W. Stephens, *Life and Letters of Edward A. Freeman*, 2 vols (London: Macmillan Company, 1895), vol. 2, p. 62.
20. C. P. Fortescue (Lord Carlingford) to E. A. Freeman, 20 October 1872, Freeman Archive, 1/7/60; and in Stephens, *Life and Letters of Edward A. Freeman*, vol. 2, p. 62.
21. Freeman to Miss A. V. Ponsonby, 21 October, 1872, ibid., vol. 2, pp. 62–3.
22. B. Ponsonby to E. A. Freeman, 24 October 1872, ibid., vol. 2, p. 63.
23. Howsam, 'Academic Discipline or Literary Genre?' p. 535.
24. See Freeman to Dean Hook, 23 December 1865, in Stephens, *Life and Letters of Edward A. Freeman*, vol. 1, p. 334.
25. On Macmillan's and Charles Longman's making of historian-authors see Howsam, *Past into Print*, chs 1–3.
26. Leslie Stephen lists these articles in *Letters of John Richard Green*, pp. 500–2. The list is updated in M. M. Bevington, *The Saturday Review 1855–1868: Representative Educated Opinion in Victorian England* (New York: AMS Press, 1966), pp. 349–50. See also A. Brundage, 'Green, John Richard (1837–1883)', *Oxford Dictionary of National Biography* (Oxford: Oxford University Press, 2004).
27. Green to W. B. Dawkins, 3 July 1860, in Stephen (ed.), *Letters of John Richard Green*, pp. 44–5. For the mythologization of this debate by Green and others see I. Hesketh, *Of Apes and Ancestors: Evolution, Christianity and the Oxford Debate* (Toronto: University of Toronto Press, 2009), chs 5–6.
28. W. Stubbs to E. A. Freeman, 12 February 1863, Stubbs Letters, MS. Eng. misc. e. 148: 134; and E. A. Freeman to James Bryce, 26 November 1865, Bryce Papers, MSS 5, X-films 8/3: 70.
29. Stubbs to Freeman, 16 August [1866], Stubbs Letters, MS. Eng. misc. e. 148:162; and Stubbs to Freeman, 5 April 1864, Stubbs Letters, MS. Eng. misc. e. 148: 128.
30. Freeman to Bryce, 19 November 1865, Bryce Papers, MSS 5, X-films 8/3: 66.
31. See, for instance, Freeman to Green, 12 November 1874, Freeman Archive, 1/8/27.

32. [J. R. Green], 'Freeman's History of the Norman Conquest', *Saturday Review*, 13 April 1867, p. 471.
33. Freeman to Green, 11 February 1872, Freeman Archive, 1/8/9.
34. Green to Freeman, December 1869, in Stephen (ed.), *Letters of John Richard Green*, p. 240.
35. J. R. Green to A. Macmillan, December 1869, Macmillan Archive, Add MS 55058: 5.
36. See Freeman to Macmillan, 24 July 1870, Macmillan Archive, Add MS 55049: 152; and Freeman to Bryce, 22 July 1870, Bryce Papers, MSS 5 X-films 8/3: 253.
37. Stubbs to Freeman, 14 September 1868, MS. Eng. misc. e. 148: 209.
38. Charles Knight, *The Popular History of England: An Illustrated History of Government from the Earliest Period to Our Own Time*, 8 vols (London: Bradbury and Evans, 1856–62), vol. 1, p. viii. On the role of the picturesque in early Victorian history-writing see R. Mitchell, *Picturing the Past: English History in Text and Image 1830–1870* (Oxford: Clarendon Press, 2000), esp. ch. 5.
39. Stubbs to Freeman, 14 September 1868, Stubbs Letters, MS. Eng. misc. e. 148: 209.
40. J. R. Green to E. A. Freeman, August 1870, in Stephen (ed.), *Letters of John Richard Green*, pp. 255–6.
41. 'Publishing Agreement for A Short History of the English People', Macmillan Archive, Add MS 55058: 68
42. Brundage, 'Green, John Richard (1837–1883)'.
43. 'Green's *Short History of the English People*', *Saturday Review*, 9 January 1875, pp. 51–3 on p. 52.
44. J. Rowley, 'Mr. Green's Short History of the English People: Is It Trustworthy?' *Fraser's Magazine*, 12 (1875), pp. 395–410, 710–24.
45. A. Macmillan to Lord Stratford de Redcliffe, 27 March 1876, in G. A. Macmillan (ed.), *Letters of Alexander Macmillan* (Glasgow: Glasgow University Press, 1908), pp. 292–3.
46. Freeman to Green, 21 September 1873, Freeman Archive, 1/8/18.
47. Freeman to Green, 16 May 1875 and 21 July 1875, Freeman Archive, 1/8/33 and 1/8/37.
48. Stephens, *Life and Letters of Edward A. Freeman*, vol. 1, p. 303.
49. E. A. Freeman, 'Green's Short History', *Pall Mall Gazette*, 18 January 1875, pp. 11–12.
50. Freeman to Green, 21 January 1875, Freeman Archive, 1/8/31.
51. [J. S. Brewer], Review of *A Short History of the English People* by J. R. Green, *Quarterly Review*, 141 (1876), pp. 285–328.
52. J. R. Green, *A Short History of the English People*, vol. 1 (1874; London: Macmillan, 1881), p. xii.
53. Freeman to Bryce, 24 August 1873, Bryce Papers, Add MS 6, X-films 8/5: 50.
54. See Freeman to Macmillan, 24 July 1870, Macmillan Archive, Add MS 55049: 154.
55. L. Howsam has nicely reconstructed the history of the Historical Course for Schools series in 'Academic Discipline or Literary Genre?' pp. 536–42.
56. On women as educators in Victorian Britain see M. J. Peterson, *Family, Love and Work in the Lives of Victorian Gentlewomen* (Indianapolis, IN: Indiana University Press, 1989), ch. 5; and P. Levine, *Feminist Lives in Victorian England: Private Roles and Public Commitment* (Oxford: Basil Blackwell, 1990), pp. 132–5.
57. D. M. Copelman, *London's Women Teachers: Gender, Class and Feminism 1870–1930* (London/New York: Routledge, 1996), p. 9.
58. See, for instance, B. Lightman, *Victorian Popularizers of Science: Designing Nature for New Audiences* (Chicago, IL: University of Chicago Press, 2007), ch. 3; and Lightman, 'Constructing Victorian Heavens: Agnes Clerk and the "New Astronomy"', in B. T.

Gates and A. B. Shteir (eds), *Natural Eloquence: Women Reinscribe Science* (Madison: University of Wisconsin Press, 1997), pp. 43–75.
59. Howsam, 'Academic Discipline or Literary Genre?' p. 536.
60. J. R. Green to E. A. Freeman, June 1870, in Stephen (ed.), *Letters of John Richard Green*, p. 254.
61. J. R. Green to E. A. Freeman, 27 June 1871, ibid., p. 305.
62. [E. A. Freeman], 'The Art of History-Making', *Saturday Review*, 17 November 1855, pp. 52–4 on p. 52.
63. J. R. Green to E. A. Freeman, 27 June 1871, in Stephen (ed.), *Letters of John Richard Green*, p. 304
64. J. R. Green to E.A. Freeman, 30 December 1872, ibid., p. 340.
65. Green to Macmillan, 12 May 1872, Macmillan Archive, Add MS 55058: 19.
66. Howsam, 'Academic Discipline or Literary Genre?' p. 538.
67. Freeman to Macmillan, 19 January 1872, Macmillan Archive, Add MS 55050: 5; and cited in Howsam, 'Academic Discipline or Literary Genre?' p. 538.
68. Stephens, *Life and Letters of Edward A. Freeman*, vol. 2, p. 463.
69. M. W. Rossiter, '"Women's Work" in Science, 1880–1910', *Isis*, 71 (1980), pp. 381–98, esp. pp. 383–7 (on 'Pickering's Harem').
70. Freeman to Green, 10 October 1873, Freeman Archive, 1/8/19.
71. Freeman to Green, 5 November 1873, Freeman Archive, 1/8/20.
72. L. Creighton, *Life and Letters of Mandell Creighton*, 2 vols (London: Longmans, Green and Company, 1904), vol. 1, p. 146.
73. Ibid., vol. 1, p. 147.
74. J. Covert, *A Victorian Marriage: Mandell and Louise Creighton* (New York/London: Hambledon and London, 2000), pp. 91–3.
75. *A Short History of the English People*, ed. A. S. Green rev. edn (New York: Harper & Brothers, 1894); A. S. Green, *Town Life in the Fifteenth Century*, 2 vols (London: Macmillan, 1894); and A. S. Green, 'Woman's Place in Literature', *Nineteenth Century*, 41 (1897), pp. 964–74. See also R. B. McDowell, *Alice Stopford Green: A Passionate Historian* (Dublin: Allen Figgis and Company, 1967).
76. Seeley to Macmillan, 7 July 1872, Macmillan Archive, Add MS 55074: 8.
77. Grove to Seeley, 13 December 1875, Macmillan Archive, Add MS 55074: 13.
78. Seeley to unknown, 12 March 1890[?], Macmillan Archive Add MS 55074: 56.
79. Mary A. P. Seeley to [?], 11 May 1901, Macmillan Archive, Add MS 55074: 80.
80. He initially mentions the new work in a letter to Macmillan dated 6 February 1877, Macmillan Archive, Add MS 55074: 18.
81. Seeley to Macmillan, 4 October [1882], quoted in L. Howsam, 'Imperial Publishing and the Idea of Colonial History, 1870–1916', *History of Intellectual Culture* (2005), pp. 1–30 on p. 6
82. Ibid., pp. 6–10; and Howsam, *Past into Print*, pp. 53–4.
83. Seeley to Macmillan, 6 February 1882, Macmillan Archive, Add MS 55074: 24.
84. J. R. Seeley, *The Expansion of England* (1883; London: Macmillan and Company, 1906), pp. 358–9.
85. 'The English Historical Review', *Saturday Review*, 23 January 1886, p. 128.
86. Bryce to Acton, 25 December 1886, Acton Papers, Add MS 8119(1): B219.
87. Creighton to Bryce, 14 August 1886, Bryce Papers, Add MS 281, X-films 8/5: 65.
88. Bryce to Acton, 25 December 1886, Acton Papers, Add MS 8119(1): B219.
89. Creighton to Bryce, 14 August 1886, Bryce Papers, Add MS 281, X-films 8/5: 65.

90. Creighton to Bryce, 28 November 1889, Bryce Papers, Add MS 53, X-films 8/5: 156. On the Royal Historical Society see J. W. Burrow, 'Victorian Historians and the Royal Historical Society', *Transactions of the Royal Historical Society*, 39 (1989), pp. 125–40.
91. Creighton to Bryce, 14 August 1886, Bryce Papers, Add MS 281, X-films 8/5: 65.
92. Creighton to Acton, 24 January 1887, Acton Papers, Add MS 8119(1): C255.
93. Creighton to W. E. Gladstone, 19 January 1887, in Creighton, *Life and Letters of Mandell Creighton*, vol. 1, p. 342.
94. Creighton to Gladstone, 15 February 1887, ibid.
95. Creighton to C. J. Longman, 21 January 1887, ibid.
96. Creighton to Longman[?] 2 Jan 1888, Bryce Papers, Add MS 53, X-films 8/5: 161.
97. Creighton to Longman, 28 October 1889, in Creighton, *Life and Letters of Mandell Creighton*, vol. 1, pp. 343–4. See also D. Goldstein, 'The Origins and Early Years of the English Historical Review', *English Historical Review*, 101 (1986), pp. 6–19 on pp. 13–14.

7 The Death of the Historian

1. John Evans, 'Address to the Society of Antiquaries', 23 April 1892, p. 12, in Freeman Archive, 8/3/5.
2. It was G. Read's paper, 'Reading Obituaries: Death, Masculinity and Republicanism in Interwar France, 1919–1940', presented at the New Frontiers in Graduate History Conference, York University, Toronto, Canada, 24 February 2004, that provided me with the impetus to analyze the obituaries of historians. See his '*Des Hommes et des citoyens*: Paternalism and Citizenship on the Republican Right in Interwar France, 1919–1940', *Historical Reflections/Réflexions Historiques*, 34 (2008), pp. 88–111. See also W. A. Stross, 'Magazines of Mortality: A Cultural History of the Obituary in Eighteenth-Century London', Ph.D. diss., University of Toronto, 2004.
3. J. Bryce, 'Edward Augustus Freeman', *English Historical Review*, 7 (1892), pp. 497–509 on p. 497.
4. B. Norton, *Freeman's Life: Highlights, Chronology, Letters and Works* (Farnborough, Norton, 1993), p. 13.
5. Bryce, 'Edward Augustus Freeman', p. 504.
6. 'Mr. E. A. Freeman', *Saturday Review*, 19 March 1892, p. 324.
7. 'Professor Freeman', *Speaker*, 10 March 1892, pp. 342–3 on p. 342.
8. 'Prof. Freeman', *Athenaeum*, 19 March 1892, p. 374.
9. 'Mr. E. A. Freeman', *Saturday Review* 19 March 1892, p. 324.
10. St. Leo Strachey, 'The Late E. A. Freeman', *Literary Opinion*, May 1892, pp. 53–7 on p. 55, in Freeman Archive, 8/3/9.
11. Bryce, 'Edward Augustus Freeman', p. 504.
12. Strachey, 'The Late E. A. Freeman', pp. 53–7.
13. Bryce, 'Edward Augustus Freeman', p. 507.
14. 'Prof. Freeman', *Athenaeum*, p. 374.
15. 'Professor Freeman', *Speaker*, p. 343.
16. Bryce, 'Edward Augustus Freeman', p. 507.
17. Ibid., p. 497–8.
18. 'Professor Edward A. Freeman', *Archaeological Journal*, 49 (1892), p. 86, in Freeman Archive, 8/3/3.

19. C. R. M[arkham], 'The Late Professor E. A. Freeman and His Services to Geography', *Proceedings of the Royal Geographical Society and Monthly Record of Geography*, 14 (June 1892), pp. 401–4, in Freeman Archive, 8/3/12.
20. Ibid., p. 403, in Freeman Archive, 8/3/12.
21. Bryce, 'Edward Augustus Freeman', p. 500.
22. Ibid., p. 498.
23. Ibid., p. 508.
24. Ibid., p. 505.
25. Ibid., pp. 505–6.
26. Strachey, 'The Late E. A. Freeman', p. 55, in Freeman Archive, 8/3/9.
27. J. E[earle] and W. A. B. C[oolidge], 'In Memoriam—E. A. Freeman', *The Guardian*, 23 March 1892, p. 442, in Freeman Archive, 8/3/1.
28. 'Professor Freeman', *Speaker*, p. 343.
29. Bryce, 'Edward Augustus Freeman', p. 507.
30. R. J. Evans, *In Defense of History* (New York: W. W. Norton and Company, 1999), pp. 139, 241 n. 1. 'Joan W. Scott criticizes an unnamed American colleagues for ascribing this famous saying to Herbert Baxter Adams, only to misattribute it herself to Edward Freeman.' Now we can add Evans to the list of those misattributing the phrase. He claims it was originally said by Seeley though it is well-established that Freeman coined the phrase (and continued to repeat it). He first does so in a lecture in Birmingham, 18 November 1880, printed in 'On the Study of History', *Fortnightly Review*, 35 (1881), pp. 319–39 on p. 320. He later repeats the phrase in 1881 at a lecture at Johns Hopkins University printed in *Lectures to American Audiences* (Philadelphia, PA: Porter & Coates, 1882), pp. 207–8. The phrase would subsequently adorn the frontispiece of the Johns Hopkins University Studies in Historical and Political Science series and it would also later be adopted by the Johns Hopkins department of history as its official motto. See H. B. Adams, 'Mr. Freeman's Visit to Baltimore', *An Introduction to American Institutional History* (Baltimore, MD: Johns Hopkins University Press, 1882), p. 12. Evans can blame the *Oxford Dictionary of Quotations* 2nd edn (Oxford: Oxford University Press, 1953) for the original misattribution of the phrase to Seeley.
31. [J. Bryce], 'Prefatory Note', *English Historical Review*, 1 (1886), p. 2. Though note as well that Bryce attributes the famous phrase properly to Freeman.
32. J. R. Tanner, 'John Robert Seeley', *The English Historical Review*, 10 (1895), pp. 507–14 on pp. 507–8.
33. 'Sir John Seeley, K.C.M.G.', *Academy*, 19 January 1895, p. 57.
34. 'Sir John Seeley', *Saturday Review*, 19 January 1895, p. 89.
35. M. Todhunter, 'Sir John Seeley', *Westminster Review*, 145 (1896), pp. 503–9 on p. 508.
36. Tanner, 'John Robert Seeley', p. 508.
37. 'Sir John Seeley, K.C.M.G.', p. 57.
38. 'Sir John Seeley', *Saturday Review*, p. 89.
39. H. A. L. Fisher, 'Sir John Seeley', *Fortnightly Review*, 60 (1896), pp. 183–99 on p. 183.
40. Tanner, 'John Robert Seeley', p. 508.
41. 'Sir John Seeley, K.C.M.G.', p. 57.
42. Todhunter, 'Sir John Seeley', p. 506.
43. Ibid., 504.
44. 'Sir John Seeley, K.C.M.G.', p. 57.
45. Fisher, 'Sir John Seeley', p. 183.
46. 'Sir John Seeley, K.C.M.G.', p. 57.

47. 'Sir John Seeley', *Saturday Review*, p. 89.
48. Tanner, 'John Robert Seeley', p. 508.
49. 'Sir John Seeley', *Saturday Review*, p. 89.
50. Todhunter, 'Sir John Seeley', p. 506.
51. Tanner, 'John Robert Seeley', p. 509.
52. Ibid., 511.
53. Ibid., p. 514.
54. F. W. Maitland to R. L. Poole, 29 April 1901, in C. H. S. Fifoot (ed.), *The Letters of Frederic William Maitland* (London: Selden Society, 1965), p. 225. 'Your request for some words about Stubbs is distressing me.' On Maitland see G. R. Elton, *F. W. Maitland* (London: Weidenfeld and Nicolson, 1986).
55. Maitland to Poole, 29 April 1901, in Fifoot (ed.), *The Letters of Frederic William Maitland*, p. 225.
56. Ibid., p. 226.
57. For an analysis of the epistemological relationship between the texts of Stubbs, Maitland and Macaulay see J. Vernon, 'Narrating the Constitution: The Discourse of "the Real" and the Fantasies of Nineteenth-Century Constitutional History', in Vernon (ed.), *Re-Reading the Constitution: New Narratives in the Political History of England's Long Nineteenth Century* (Cambridge: Cambridge University Press, 1996), pp. 204–29. For the ways in which Maitland's interpretation of the constitution subtly subverted Stubbs see M. Bentley, *Modernizing England's Past: English Historiography in the Age of Modernism, 1870–1970* (Cambridge: Cambridge University Press, 2005), pp. 32–5. On the relationship between Stubbs and Maitland and to German scholarship see J. Campbell, 'Stubbs, Maitland, and Constitutional History', in B. Stuchtey and P. Wende (eds), *British and German Historiography 1750–1950: Traditions, Perceptions, and Transfers* (Oxford: Oxford University Press, 2000), pp. 99–122.
58. F. W. Maitland, 'William Stubbs, Bishop of Oxford', *The English Historical Review*, 16 (1901), p. 417.
59. 'The Late Bishop of Oxford', *Athenaeum*, 27 April 1901, pp. 530–1 on p. 530.
60. 'Stubbs, Dr., Late Bishop of Oxford', *Academy*, 27 April 1901, p. 356.
61. 'The Late Bishop of Oxford', *Athenaeum*, p. 351.
62. Quoted in 'The Literary Week', *Academy*, 18 May 1901, p. 415.
63. Maitland, 'William Stubbs, Bishop of Oxford', p. 422.
64. For the centrality of Stubbs's *Constitutional History* in the training of historians in the second half of the nineteenth century see R. Soffer, *Discipline and Power: The University, History, and the Making of an English Elite, 1870–1930* (Stanford, CA: Stanford University Press, 1994), pp. 86–90; and Bentley, *Modernizing England's Past*, pp. 23–4.
65. Campbell, 'Stubbs, Maitland, and Constitutional History', p. 102.
66. J. W. Burrow, *A Liberal Descent: Victorian Historians and the English Past* (Cambridge: Cambridge University Press, 1981), pp. 129–30.
67. 'Stubbs, Dr., Late Bishop of Oxford', *Academy*, p. 356.
68. 'The Late Bishop of Oxford', *Athenaeum*, p. 530.
69. Stubbs, 'Inaugural', pp. 1, 19. On Stubbs's founding of the Oxford School see D. S. Goldstein, 'The Professionalization of History in Britain in the Late Nineteenth and Early Twentieth Centuries', *Storia della Storiografia*, 3 (1983), pp. 9–14; Goldstein, 'History at Oxford and Cambridge: Professionalization and the Influence of Ranke', in G. G. Iggers and J. M. Powell (eds), *Leopold von Ranke and the Shaping of the Historical Discipline* (Syracuse, NY: Syracuse University Press, 1990), pp. 142–5; and P. R. H. Slee, *Learning*

and a Liberal Education: The Study of Modern History in the Universities of Oxford, Cambridge, and Manchester, 1840–1914* (Manchester: Manchester University Press, 1986), pp. 86–90.
70. Stubbs, *The Constitutional History of England*, vol. 1, p. v; and see Campbell, 'Stubbs, Maitland, and Constitutional History', p. 105.
71. 'Stubbs, Dr., Late Bishop of Oxford', *Academy*, p. 356.
72. 'The Late Bishop of Oxford', *Athenaeum*, p. 531.
73. Maitland, 'William Stubbs, Bishop of Oxford', pp. 421–2.
74. P. Levine, *The Amateur and the Professional: Antiquarians, Historians and Archaeologists in Victorian England, 1838–1886* (Cambridge: Cambridge University Press, 1986), p. 133. On Stubbs's editing skills and their influence on his *Constitutional History* see Burrow, *A Liberal Descent*, p. 135.
75. 'The Late Bishop of Oxford', *Athenaeum*, p. 530.
76. Maitland, 'William Stubbs, Bishop of Oxford', pp. 419, 420.
77. Ibid., p. 423.
78. Quoted in 'The Literary Week', *Academy*, p. 415.
79. Maitland, 'William Stubbs, Bishop of Oxford', p. 419.
80. Maitland's interpretation of Stubbs's writing reminds one of Robert Boyle's attempt to create 'virtual witnesses' out of the readers of his experiments as portrayed by Shapin and Schaffer, *Leviathan and the Air-Pump*, pp. 60–5, *passim*. Not everyone was happy with this style of writing, however. Alexander Macmillan criticized Edward A. Freeman for precisely the kind of 'behind the scenes' approach Maitland praised in Stubbs. Upon reading the first draft of Freeman's *Norman Conquest* Macmillan wrote on 18 February 1867 that 'when one sits down to read a history one wants that it shall mainly be narrative ... I felt as if a great deal that was written down in your book were processes through which *your own mind* must have gone, but which necessarily can be appreciated by, a very small number of readers. ... A man does not go through the kitchens with his cook, or through the dissecting room with his physician. He wants his dinner cooked & his pill or draft in their complete form without the din of pestle or the spatter of the spit. Had Gibbon written in this style his book would have been ten times as big and not a tenth as valuable.' Quoted in L. Howsam, *Past into Print: The Publishing of History in Britain, 1850–1950* (London: The British Library; Toronto: University of Toronto Press, 2009), p. 32.
81. 'The Late Bishop of Oxford', *Athenaeum*, p. 531.
82. The *EHR* published an obituary on Creighton as well, but there was not enough space to analyze it here. See R. Garnett, 'Mandell Creighton, Bishop of London', 16 (1901), pp. 211–18.
83. As G. Levine argues in *Dying to Know: Scientific Epistemology and Narrative in Victorian England* (Chicago, IL: University of Chicago Press, 2002), the type of objectivity promoted by nineteenth-century natural philosophers and scientists, could only be achieved by transcending the limits of subjectivity, by eliminating all remnants of the self in the narrative. This method paradoxically suggests that knowledge can only be acquired in death, perhaps a variant on Hegel's owl of Minerva flying only at dusk. (I must thank D. Woolf for the reference to Minerva.)
84. Maitland to Poole, 22 June 1902, in Fifoot (ed.), *The Letters of Frederic William Maitland*, p. 246.
85. Acton Archive, Add MS 7472.
86. Lord Acton to R. Lane Poole, 30 August 1892, Acton Archive, Add MS 7472: 19.

87. Acton to Poole, 23 October 1896, Acton Archive, Add MS 7472: 24. See also Acton to Poole, 30 October 1896, Acton Archive, Add MS 7472: 27.
88. R. L. Poole, 'John Emerich, Lord Acton', *English Historical Review*, 17 (1902), pp. 692–9 on p. 692.
89. Ibid., p. 695.
90. G. P. Gooch, 'Lord Acton', *Speaker*, 28 June 1902, pp. 335–6 on 335.
91. Poole, 'John Emerich, Lord Acton', p. 693.
92. 'The Literary Week', *Academy*, 28 June 1902, p. 3.
93. 'Lord Acton', *Athenaeum*, 5 July 1902, p. 32.
94. 'The Late Lord Acton', *Edinburgh Review*, 197 (1903), pp. 501–34 on pp. 528.
95. Poole, 'John Emerich, Lord Acton', p. 94.
96. 'The Late Lord Acton', *Edinburgh Review*, pp. 503–4.
97. H. Butterfield, *Man On His Past: The Study of the History of Historical Scholarship* (Cambridge: Cambridge University Press, 1969), pp. 62–95 on p. 63.
98. Gooch, 'Lord Acton', pp. 336; and 'Lord Acton', *Athenaeum*, 28 June 1902, p. 817.
99. F. W. Maitland, 'Lord Acton', in *The Collected Papers of Frederic William Maitland*, ed. H. A. L. Fisher, vol. 3 (Cambridge: Cambridge University Press, 1911), pp. 512–21 on p. 515; orig. pub. in *The Cambridge Review*, 16 October 1902.
100. Gooch, 'Lord Acton', pp. 336.
101. Poole, 'John Emerich, Lord Acton', p. 693.
102. Gooch, 'Lord Acton', p. 336.
103. 'Lord Acton', *Athenaeum*, 5 July 1902, p. 32.
104. 'The Literary Week', *Academy*, p. 3.
105. Gooch, 'Lord Acton', p. 336.
106. 'The Late Lord Acton', *Edinburgh Review*, pp. 531–2.
107. Ibid., pp. 531.
108. 'Lord Acton', *Athenaeum*, 5 July 1902, p. 32.
109. Poole, 'John Emerich, Lord Acton', p. 694.
110. Maitland to Poole, 19 October 1902, in Fifoot (ed.), *The Letters of Frederic William Maitland*, p. 267.
111. 'The Late Lord Acton', *Edinburgh Review*, pp. 534.
112. Quoted in H. A. L. Fisher, *James Bryce*, 2 vols (London: The Macmillan Company, 1927), vol. 1, p. 195.

Epilogue: Froude's Revenge

1. A. Lang, 'History as She Ought to be Wrote', *Blackwood's Edinburgh Magazine*, 166 (August 1899), pp. 266–74 on p. 273.
2. It is important to note here that while the influence of Buckle's work steadily declined in England, his work was immensely influential in many other parts of the world well into the twentieth century, particularly in Latin America, China and Japan. See esp. Woolf, *A Global History of History*, ch. 8; and A. Schneider and S. Tanaka, 'The Transformation of History in China and Japan' in S. Macintyre, J. Maiguashca and A. Pók (eds), *The Oxford History of Historical Writing*, vol. 4: *1800–1945* (Oxford: Oxford University Press, forthcoming, 2011).
3. L. Strachey and A. Pollard both sought to rescue Froude's histories for posterity. See Strachey, 'One of the Victorians', *Saturday Review of Literature*, 6 December 1930, pp. 418–19; and [Pollard], 'Froude', *Times Literary Supplement*, 18 April 1918, pp. 177–8.

Pollard argued that '[o]nly Freeman's ferocious pedantry and the ignorance of his readers allowed Froude's peccadilloes [inaccuracies] ... to prejudice his "History". Nor is there a shadow of doubt as to Froude's industry and honesty as an historian'. Ibid., p. 178. On the posthumous vindication of Froude see Michael Bentley, *Modernizing England's Past: English Historiography in the Age of Modernism, 1870–1970* (Cambridge: Cambridge University Press, 2005), pp. 102, 110.
4. A. Jessop, Review of *Divorce of Catherine of Aragon* by J. A. Froude, *English Historical Review*, 7 (1892), pp. 360–5 on p. 360.
5. Ibid., p. 365.
6. M. A. S. Hume, Review of *The Story of the Spanish Armada and Other Essays* by J. A. Froude, *English Historical Review*, 7 (1892), pp. 567–71 on p. 567.
7. Ibid., p. 570.
8. On Froude's appointment see H. Paul, *The Life of Froude* (London: Sir Isaac Pitman and Sons Ltd., 1905), ch. 10; and J. Markus, *J. Anthony Froude: The Last Undiscovered Great Victorian* (New York: Scribner, 2005), pp. 284–8.
9. James Bryce to Lord Acton, April 1892, Bryce Papers, MSS 1/56.
10. C. H. Firth to T. F. Tout, [April?] 1892, Tout Correspondence, 1/367/29.
11. R. L. Poole to T. F. Tout, 30 April 1892, 1/953/25, Tout Correspondence, 1/953/25.
12. J. A. Froude, 'Inaugural Lecture', *Longman's Magazine*, 21 (December 1892), pp. 140–62 on p. 140.
13. Ibid., p. 144.
14. Ibid., p. 153.
15. Ibid., p. 157.
16. Ibid., p. 158.
17. Ibid., p. 161.
18. 'James Anthony Froude', *English Illustrated Magazine*, 106 (1892), pp. 721–2 on p. 722.
19. 'Mr. Froude's Appointment', *Saturday Review*, 9 April 1892, p. 411.
20. A. Jessop, Review of *Life and Letters of Erasmus* by J. A. Froude, *English Historical Review*, 10 (1895), pp. 574–6 on p. 574.
21. H. A. L. Fisher, 'Modern Historians and Their Methods', *Fortnightly Review*, 56 (1894), p. 806.
22. Ibid., p. 804.
23. Ibid., p. 805.
24. Ibid., pp. 808–9.
25. Ibid., p. 810.
26. Ibid., p. 811.
27. 'Will Froude Live?' *Speaker*, 27 October 1894, pp. 454–5 on p. 455.
28. 'James Anthony Froude', *Academy*, 27 October 1894, p. 329.
29. 'James Anthony Froude', *Saturday Review*, 27 October 1894, pp. 453–4 on p. 453.
30. Ibid., p. 454.
31. J. B. Bury, 'The Science of History', in *Selected Essays*, ed. Harold Temperly (Cambridge: Cambridge University Press, 1930), p. 3.
32. Ibid., p. 4.
33. Ibid., p. 22.
34. H. Paul, 'Professor Bury's Ideas of History', *Speaker*, 6 June 1903, p. 226–7 on p. 226.
35. Ibid., p. 227.
36. Lord Acton to R. Lane Poole, 22 January 1897, Acton Papers, Add MS 7472: 33.

37. G. M. Trevelyan, 'The Latest View of History', *Independent Review*, 1 (1903), pp. 395–414 on pp. 396–7.
38. R. N. Soffer, *Discipline and Power: The University, History, and the Making of an English Elite, 1870–1930* (Stanford, CA: Stanford University Press, 1994), p. 58.
39. On these and other issues concerning the historical practice in the twentieth century see Novick, *That Noble Dream*; H. White, *The Content of the Form: Narrative Discourse and Historical Representation* (Baltimore, MD: Johns Hopkins University Press, 1987); D. R. Kelley, *Frontiers of History: Historical Inquiry in the Twentieth Century* (New Haven, CT: Yale University Press, 2006); and A. Schneider and D. Woolf (eds), *The Oxford History of Historical Writing*, vol. 5: *Historical Writing since 1945* (Oxford: Oxford University Press, forthcoming, 2011).
40. H. White, 'The Burden of History', *History and Theory*, 5 (1866), pp. 111–34; also published in White, *Tropics of Discourse: Essays in Cultural Criticism* (Baltimore, MD: Johns Hopkins University Press, 1985), pp. 27–50. See also D. Leeson, 'Cutting Through History: Hayden White, William S. Burroughs, and Surrealistic Battle Narratives', *Left History*, 10 (2004), pp. 13–43.
41. D. P. Miller, 'The "Sobel Effect": The Amazing Tale of How Multitudes of Popular Writers Pinched All the Best Stories in the History of Science and Became Rich and Famous while Historians Languished in Accustomed Poverty and Obscurity, and How this Transformed the World', *Metascience* (2002), pp. 185–200; and I. Hesketh, 'The Sobel Effect, Froude's Disease, and the Making of Un-Popular History in Mid-Victorian England,' paper presented at the *History of Science Society Annual Conference*, November 2006, Vancouver, BC.

WORKS CITED

Manuscripts

Acton Papers: The Papers of Lord Acton, University Library, Cambridge University, UK.

Buckle Collection: H. T. Buckle Collection, University of Illinois at Urbana-Champaign, USA.

Bryce Papers: The Papers of James Bryce, Bodleian Library, Oxford, UK.

Freeman Archive: Edward A. Freeman Archive, John Rylands Library of Manchester, UK.

Macmillan Archive: Alexander Macmillan Archive, British Library, London UK.

Seeley Papers: The Papers of J. R. Seeley, University of London, UK.

Stubbs Letters: William Stubbs to E. A. Freeman, MS. Eng. misc. e. 148, Bodleian Library, Oxford University, UK.

Tout Correspondence: T. F. Tout Correspondence, John Rylands Library of Manchester, UK.

Primary Sources

Acton, Lord, 'The Catholic Press', *Rambler*, 11 (February 1859), pp. 73–90.

—, 'Mr. Goldwin Smith's *Irish History*', *Rambler*, 6 (January 1862), pp. 190–229.

—, Review of *A History of the Papacy during the Period of the Reformation* by Mandell Creighton, *Academy*, 9 December 1882, pp. 407–9.

—, 'German Schools of History', *The English Historical Review*, 1 (1886), pp. 7–43.

—, Review of *A History of Papacy During the Period of the Reformation* by Mandell Creighton, vols 3 and 4, *The English Historical Review*, 2 (1887), pp. 571–81.

—, 'Döllinger's Historical Work', *English Historical Review*, 5 (1890), pp. 700–44.

—, *Letters of Lord Acton to Mary Gladstone*, ed. H. Paul (London: The Macmillan Company, 1904).

—, *Lectures on Modern History*, ed. N. Figgis and R. Laurence (London: Macmillan and Company, 1906).

—, *The History of Freedom and other Essays*, ed. N. Figgis and R. V. Laurence (London: Macmillan and Co., 1922).

—, *Essays in the Liberal Interpretation of History: Selected Papers*, ed. W. H. McNeill (Chicago, IL: University of Chicago Press, 1967).

—, *The Correspondence of Lord Acton and Richard Simpson*, ed. J. L. Altholz and D. McElrath, vol. 1 (Cambridge: Cambridge University Press, 1971).

—, *Selected Writings of Lord Acton*, vol. 2: *Essays in the Study and Writing of History*, ed. J. R. Fears (Indianapolis, IN: Liberty Fund, 1986).

Adams, H. B., 'Mr. Freeman's Visit to Baltimore', *An Introduction to American Institutional History* (Baltimore, MD: Johns Hopkins University Press, 1882).

Anon., 'Mr. Kingsley on the Science of History', *The Literary Gazette*, 15 December 1860, pp. 509–10.

—, 'Mr. Kingsley on Science and History', *London Review*, 22 December 1860, pp. 599–600.

—, 'The Oxford Professoriate', *Saturday Review*, 12 July 1862, p. 46.

—, Review of *History of England* by James Rawson Gardiner, *Athenaeum*, 21 March 1863, pp. 392–3.

—, 'Mr. Froude on the Science of History', *The Reader*, 27 February 1864, pp. 255–6.

—, 'Ecce Homo', *Spectator*, 23 December 1865, p. 1436.

—, 'Literary Gossip', *London Review*, 17 November 1866, pp. 559–560.

—, 'Mr. Froude's Short Studies on Great Subjects', *London Review*, 6 April 1867, pp. 406–7.

—, 'Our Cambridge Letter', *Athenaeum*, 26 February 1870, p. 291.

—, 'Seeley's Lectures and Essays', *Saturday Review*, 7 January 1871, pp. 23–4.

—, 'Seeley's Life and Times of Stein', *Saturday Review*, 1 February 1879, pp. 146–7.

—, 'Some Papers in the Magazines', *Spectator*, 5 April 1879, pp. 442–4.

—, 'The English Historical Review', *Saturday Review*, 23 January 1886, p. 128.

—, 'James Anthony Froude', *English Illustrated Magazine*, 106 (1892), pp. 721–2.

—, 'Professor Edward A. Freeman', *Archaeological Journal*, 49 (1892), p. 86.

—, 'Professor Freeman', *Speaker*, 10 March 1892, pp. 342–3.

—, 'Mr. E. A. Freeman', *Saturday Review*, 19 March 1892, p. 324.

—, 'Prof. Freeman', *Athenaeum*, 19 March 1892, p. 374.

—, 'Mr. Froude's Appointment', *Saturday Review*, 9 April 1892, p. 411.

—, 'James Anthony Froude', *Academy*, 27 October 1894, p. 329.

—, 'James Anthony Froude', *Saturday Review*, 27 October 1894, pp. 453–4.

—, 'Will Froude Live?' *Speaker*, 27 October 1894, pp. 454–5.

—, 'Sir John Seeley, K.C.M.G.', *Academy*, 19 January, 1895, p. 57.

—, 'Sir John Seeley', *Saturday Review*, 19 January 1895, p. 89.

—, 'The Late Bishop of Oxford', *Athenaeum*, 27 April 1901, pp. 530–1.

—, 'Stubbs, Dr., Late Bishop of Oxford', *Academy*, 27 April 1901, p. 356.

—, 'The Literary Week', *Academy*, 18 May 1901, p. 415.

—, 'The Literary Week', *Academy*, 28 June 1902, p. 3.

—, 'Lord Acton', *Athenaeum*, 28 June 1902, p. 817.

—, 'Lord Acton', *Athenaeum*, 5 July 1902, p. 32.

—, 'The Late Lord Acton', *Edinburgh Review*, 197 (1903), pp. 501–34.

—, 'Lord Acton's Inaugural Lecture', *Speaker*, 15 June 1895, pp. 648–9.

—, 'Lord Acton on Trial', *Saturday Review*, 22 June 1895, pp. 821–2.

[Beesly, E. S.], 'Mr. Kingsley on the Study of History', *Westminster Review*, 75 (1861), pp. 305–36.

[Brewer, J. S.], Review of *A Short History of the English People* by J. R. Green, *Quarterly Review*, 141 (1876), pp. 285–328.

Brewster, D., *The Life of Sir Isaac Newton* (1831; New York: Harper, 1840).

[Bryce, J.], 'Prefatory Note', *The English Historical Review*, 1 (1886), pp. 1–6.

—, 'Edward Augustus Freeman', *The English Historical Review*, 7 (1892), pp. 497–509.

Buckle, H. T., *History of Civilization in England*, 2 vols (1857, 1861; New York: D. Appleton and Company, 1860, 1861).

—, 'The Influence of Women on the Progress of Knowledge', *Fraser's Magazine*, 57 (April 1858), pp. 395–407.

—, 'Mill on Liberty', *Fraser's Magazine*, 59 (May 1859), pp. 509–42.

—, *Miscellaneous and Posthumous Works of Henry Thomas Buckle*, ed. H. Taylor, 3 vols (London: Longmans, Green, and Co., 1872).

Bury, J. B., *Selected Essays*, ed. H. Temperly (Cambridge: Cambridge University Press, 1930).

Burrows, M., *Autobiography of Montagu Burrows*, ed. S. M. Burrows (London: Macmillan and Company, 1908).

[Carlyle, T.], 'Thoughts on History', *Fraser's Magazine*, 10:2 (November 1830), pp. 413–18.

Coleridge, J. D., 'A letter to the Editor from Mr. J. D. Coleridge', *Fraser's Magazine*, 59 (1859), pp. 635–6.

Comte, A., 'The Positive Philosophy and the Study of Society', in *Theories of History*, ed. P. Gardiner (New York: The Free Press, 1959), pp. 75–9.

[Cordery, B. M.], 'Gardiner's *Reign of James I*', *Edinburgh Review*, 143:291 (January 1876), pp. 101–41.

Creighton, L., *The Life and Letters of Mandell Creighton*, 2 vols (London: Longmans, Green, and Co., 1905).

Creighton, M., 'The Endowment of Research', *Macmillan's Magazine*, 34 (1876), pp. 186–92.

—, 'Introductory Note', in Ward, Prothero, and Leathes (eds), *The Cambridge Modern History*, vol. 1, pp. 1–6.

—, 'Historical Ethics', *Quarterly Review* (1905), pp. 32–46.

Darwin, C., *On the Origin of Species*, 6th edn (1872; New York Mentor, 1958).

—, *The Autobiography of Charles Darwin, 1809–1882*, ed. N. Barlow (New York: W. W. Norton and Company, 1969).

—, *The Correspondence of Charles Darwin*, vol. 7: *1858–1859*, ed. F. B. and S. Smith (Cambridge: Cambridge University Press, 1991).

Darwin, F. (ed.), *The Life and Letters of Charles Darwin*, 2 vols (London: D. Appleton 1887).

Dickens, C., *Hard Times* (London, 1854).

Droysen, J. G., *Outline of the Principles of History*, trans. E. B. Andrews (Boston, MA: Ginn and Company, 1893); orig. pub as 'Die Erhebung der Geschichte zum Rang einer Wissenschaft', *Historische Zeitschrift*, 9 (1862), pp. 1–22.

E[earle], J. and W. A. B. C[oolidge], 'In Memoriam—E. A. Freeman', *The Guardian*, 23 March 1892, p. 442.

[Eliot, G.], Review of *The Progress of the Intellect, as Exemplified in the Religious Development of the Greeks and Hebrews* by Robert William Mackay, *Westminster Review*, 54:2 (January 1851), pp. 353–68.

[—], 'Rationalism in Europe', *Westminster Review* 28:2 (October 1865), pp. 326–51.

[Elwin, W.], 'Ecce Homo', *Quarterly Review*, 119 (1866), pp. 515–29.

Fisher, H. A. L., 'Modern Historians and Their Methods', *Fortnightly Review*, 56 (1894), pp. 803–16.

—, 'Sir John Seeley', *Fortnightly Review*, 60 (1896), pp. 183–99.

—, *James Bryce*, 2 vols (London: The Macmillan Company, 1927).

[Freeman, E. A.], 'The Art of History-Making', *Saturday Review*, 17 November 1855, pp. 52–4.

[—], Review of *Revolutions in English History* by Dr. Vaughan, *Edinburgh Review*, 112 (1860), pp. 136–60.

[—], 'Froude's Reign of Elizabeth (First Notice)', *Saturday Review*, 16 January 1864, pp. 80–2.

[—], 'Mr. Kingsley's Roman and Teuton', *Saturday Review*, 9 April 1864, pp. 446–8.

[—], 'The Mythical and Romantic Elements in Early English History', *The Fortnightly Review*, 24 (1866), pp. 641–68.

[—], 'Kingsley's Hereward', *Saturday Review*, 19 May 1866, pp. 594–5.

[—], 'Professor Kingsley on the "Ancien Regime"', *Saturday Review*, 22 June 1867, pp. 792–3.

—, *The History of the Norman Conquest of England, its Causes and its Results*, vol. 1, 2nd edn (1867; London: Macmillan, 1870).

—, *Old English History for Children* (London: Macmillan, 1869).

—, *Comparative Politics* (London: Macmillan and Co., 1874).

—, 'Green's Short History', *Pall Mall Gazette*, 18 January 1875, pp. 11–12.

—, 'Mr. Froude's Life and Times of Thomas Becket (Part I)', *Contemporary Review*, 31 (1878), pp. 821–42.

—, 'Mr. Froude's Life and Times of Thomas Becket (Part II)', *The Contemporary Review*, 32 (1878), pp. 116–39.

—, 'Mr. Froude's Life and Times of Thomas Becket (Part III)', *The Contemporary Review*, 32 (1878), pp. 474–501.

—, 'Mr. Froude's Life and Times of Thomas Becket (Part IV)', *The Contemporary Review*, 33 (1878), pp. 213–41.

—, 'Last Words on Mr. Froude', *Contemporary Review*, 35 (1879), pp. 215–36.

—, 'On the Study of History', *Fortnightly Review*, 35 (1881), pp. 320–39.

—, *Lectures to American Audiences* (Philadelphia, PA: Porter & Coates, 1882).

—, 'John Richard Green', *British Quarterly Review*, 155 (1883), pp. 69–74.

—, 'The Tyrants of Britain, Gaul, and Spain, A.D. 406–411', *English Historical Review*, 1 (1886), pp. 53–85.

—, *The Methods of Historical Study* (London: The Macmillan Company, 1886).

—, *Historical Essays*, vol. 1, 5th edn (1871; London: Macmillan and Co., 1896).

Froude, J. A., *The Nemesis of Faith* (London: John Chapman, 1849).

[—], 'England's Forgotten Worthies', *Westminster Review*, 58 (1852), pp. 32–67.

—, *The History of England from the Death of Cardinal Wolsey to the Defeat of the Spanish Armada*, 12 vols. (London: Longmans, Green, and Co., 1856–70).

—, 'A Few Words on Mr. Freeman', *Nineteenth Century*, 5 (1879), pp. 618–37.

—, *Thomas Carlyle*, 2 vols (London: Longmans, Green, 1882–4).

—, 'Inaugural Lecture', *Longman's Magazine*, 21 (December 1892), pp. 140–62.

—, *Short Studies on Great Subjects*, 4 vols (London: Longmans, Green, and Co., 1898).

Froude, R. H., *Remains of the Late Reverend Richard Hurrell Froude*, ed. J. H. Newman and J. Keble, 4 vols (London: J. G. & F. Rivington, 1838).

Garnett, R., 'Mandell Creighton, Bishop of London', *The English Historical Review*, 16 (1901), pp. 211–8.

Gladstone, W. E., *On Ecce Homo* (London: Strahan & Co. Publishers, 1868).

Gooch, G. P., 'Lord Acton', *Speaker*, 28 June 1902, pp. 335–6.

Green, A. S., *Town Life in the Fifteenth Century*, 2 vols. (London: Macmillan, 1894).

—, 'Woman's Place in Literature', *Nineteenth Century*, 41 (1897), 964–74.

Green, J. R., *A Short History of the English People* (1874; London: Macmillan, 1881).

[—], 'Professor Stubbs's Inaugural Lecture', *Saturday Review*, 2 March 1867, p. 279.

[—], 'Freeman's Norman Conquest', *Saturday Review*, 13 September 1867, p. 471.

—, *A Short History of the English People*, ed. A. S. Green, rev. edn (New York: Harper & Brothers, 1894).

—, *Letters of John Richard Green*, ed. L. Stephen (New York: The Macmillan Company, 1901).

Harrison, F., *The Meaning of History and Other Historical Pieces* (London: Macmillan and Company, 1921).

Holyoake, G. J., *Sixty Years of an Agitator's Life*, 2 vols (London: T. Fisher Unwin, 1892).

[Hood, E. P.], 'The Roman and the Teuton', *Eclectic Review*, 7 (July 1864), pp. 82–8.

Hume, M. A. S., Review of *The Story of the Spanish Armada and Other Essays* by J. A. Froude, *English Historical Review*, 7 (1892), pp. 567–71.

Huth, A. H., *The Life and Writings of Henry Thomas Buckle* (London: Sampson Low, Marston, Searle Rivington, 1880).

Jessop, A., Review of *Divorce of Catherine of Aragon* by J. A. Froude, *English Historical Review*, 7 (1892), pp. 360–5.

—, Review of *Life and Letters of Erasmus* by James Anthony Froude, *English Historical Review*, 10 (1895), pp. 574–6.

Kingsley, C., *Alton Locke, Tailor Poet*, 3rd edn (1850; London: Chapman and Hall, 1852).

—, *The Roman and the Teuton* (London: Macmillan and Co., 1864).

—, *Literary and General Lectures and Essays* (London: Macmillan, 1890).

Knight, C., *The Popular History of England: An Illustrated History of Society and Government from the Earliest Period to Our Own Times*, 8 vols (London: Bradbury and Evans, 1856–62).

Lang, A., 'History as She Ought to be Wrote', *Blackwood's Edinburgh Magazine*, 166 (August 1899), pp. 266–74.

Langlois, C. V. and C. Seignobos, *Introduction to the Study of History*, trans. G. G. Berry (New York: Henry Holt and Company, 1898).

[Lawrance, H.], Review of *The Limits of Exact Science as applied to History* by Charles Kingsley, *The British Quarterly Review*, 33 (1861), pp. 264–7.

Lecky, W. E. H., *History of the Rise and Influence of the Spirit of Rationalism in Europe*, 2 vols (London: Longmans, 1865).

[Lester, J. D.], 'History and Biography', *Westminster Review*, 43 (1873), pp. 301–18.

Lilly, W. L., 'The New Spirit in History', *Nineteenth Century*, 38 (1895), pp. 619–633.

[Macaulay, T. B.], 'Lord Bacon', *Edinburgh Review*, 65 (1837), pp. 3–104.

Macmillan, A., *Letters of Alexander Macmillan*, ed. G. A. Macmillan (Glasgow: Glasgow University Press, 1908).

Maitland, F. W., 'William Stubbs, Bishop of Oxford', *The English Historical Review*, 16 (1901), pp. 417–26.

—, 'Lord Acton' in *The Collected Papers of Frederic William Maitland*, ed. H. A. L. Fisher, vol. 3 (Cambridge: Cambridge University Press, 1911), pp. 512–21; orig. pub. in *The Cambridge Review*, 16 October 1902.

— (ed.), *The Life and Letters of Leslie Stephen* (London: Duckworth and Co. 1906).

—, *The Letters of Frederic William Maitland*, ed. C. H. S. Fifoot (London: Selden Society, 1965).

M[arkham], C. R., 'The Late Professor E. A. Freeman and His Services to Geography', *Proceedings of the Royal Geographical Society and Monthly Record of Geography*, 14 (1892), pp. 401–4.

McCabe, J., *The Life and Letters of George Jacob Holyoake*, 2 vols (London: Watts & Co., 1908).

[Mill, J. S.], 'Bentham', *The London and Westminster Review*, 31 (August 1838), pp. 467–506.

[—], 'Coleridge', *The London and Westminster Review*, 33 (March 1840), pp. 257–302.

—, *A System of Logic*, 8th edn (1843; London: Longmans, 1967).

—, *Principles of Political Economy*, vol. 1 (London, 1849).

Morley, J., 'Mr. Froude on the Science of History', *Fortnightly Review*, 2 (1867), pp. 226–37.

Müller, F. M., *Lectures on the Science of Language* (London: Longman, Green, Longman & Roberts, 1861).

—, 'Preface' in C. Kingsley, *The Roman and the Teuton: A Series of Lectures*, 2nd edn (London: Macmillan and Company, 1891).

—, *Auld Lang Syne* (London: C. Scribner's Sons, 1898).

Paul, H., 'Professor Bury's Ideas of History', *Speaker*, 6 June 1903, pp. 226–7.

[Pattison, M.], 'Mackay's Tübingen School', *Westminster Review*, 24:2 (October 1863), pp. 510–31.

[Pollard, A.], 'Froude', *Times Literary Supplement*, 18 April 1918, pp. 177–8.

Poole, R. L., 'John Emerich, Lord Acton', *The English Historical Review*, 17 (1902), pp. 692–9.

Powell, B., 'On the Study of the Evidences of Christianity' (1860), in *Essays and Reviews: The 1860 Text and Its Reading*, ed. V. Shea and W. Whitla (Charlottesville, VA and London: University of Virginia Press, 2000).

Powell, F. Y., 'To the Reader' in C. V. Langlois and C. Seignobos, *Introduction to the Study of History*, trans. G. G. Berry (New York: Henry Holt and Company, 1898).

Ranke, L. von., 'Preface: *Histories of the Latin and Germanic Nations from 1494–1514*, reprinted in *The Varieties of History from Voltaire to the Present*, ed. F. Stern (New York: Vintage Books, 1973), p. 57.

—, *The Theory and Practice of History*, trans. W. A. Iggers and K. von Moltke; ed. G. G. Iggers and K. von Moltke (Indianapolis, IN: The Bobbs-Merrill Company, 1973).

Robertson, J. M., *Buckle and His Critics: A Study in Sociology* (London: S. Sonnenschein, 1895).

Ross, J. (ed.), *Three Generations of English Women: Memoirs and Correspondence of Mrs. John Taylor, Mrs. Sarah Austin, and Lady Duff Gordon*, 3 vols (London: John Murray, 1888).

Rowley, J., Review of *A Short History of the English People* by J. R. Green, *Fraser's Magazine*, 12 (1875), pp. 395–410, 710–24.

[Sanders, T. C.], 'Two Years Ago', *Saturday Review*, 21 February 1857, pp. 176–7.

Seeley, J. R., *Ecce Homo: A Survey of the Life and Work of Jesus Christ* (1865; London: Macmillan, 1912).

—, *Lectures and Essays* (London: Macmillan 1870).

—, 'History and Politics I', *Macmillan's Magazine*, 40 (1879), pp. 289–99.

—, 'History and Politics II', *Macmillan's Magazine*, 40 (1879), pp. 369–78.

—, 'History and Politics III', *Macmillan's Magazine*, 40 (1879), pp. 449–58.

—, 'History and Politics IV', *Macmillan's Magazine*, 41 (1879), pp. 23–31.

—, 'Political Somnambulism', *Macmillan's Magazine*, 43 (1880), pp. 28–32.

—, *The Expansion of England* (1883; London: Macmillan and Company, 1906).

[Smith, G.], Review of *History of England from the Fall of Wolsey to the Death of Elizabeth* by J. A. Froude, vols 1–4, *Edinburgh Review*, 108 (1858), pp. 206–52.

Stephen, L., 'An Attempted Philosophy of History', *Fortnightly Review*, 27 (1880), pp. 672–95.

[—], 'Buckle, Henry Thomas (1821–1862)', in L. Stephen and S. Lee (eds), *The Dictionary of National Biography*, vol. 3 (Oxford: Oxford University Press, 1921).

—, *Some Early Impressions* (London: Leonard & Virginia Woolf at the Hogarth Press, 1924).

Stephens, W. R. W., *The Life and Letters of Edward A. Freeman*, 2 vols (London: Macmillan Company, 1895).

Strachey, L, 'One of the Victorians', *Saturday Review of Literature*, 6 December 1930, pp. 418–19; rept. as 'Froude', in *Portraits in Miniature and Other Essays* (London: Chatto & Windus, 1931), pp. 195–206.

Strachey, St. L., 'The Late E. A. Freeman', *Literary Opinion*, May 1892, pp. 53–7.

Stubbs, W., *The Constitutional History of England in Its Origin and Development*, vol. 1, 4th edn (1873; Oxford: Clarendon Press, 1883).

—, *The Constitutional History of England in Its Origins and Development*, vol. 3 (1878; Oxford: Clarendon Press, 1880).

—, *The Letters of William Stubbs, Bishop of Oxford 1825–1901*, ed. W. H. Hutton (London: Archibald Constable and Co., 1904).

—, *Lectures on Early English History*, ed. A. Hassall (London: Longmans, Green, and Co., 1906).

—, *Seventeen Lectures on the Study of Medieval and Modern History and Kindred Subjects* (New York: Howard Fertig, 1967).

Tanner, J. R., 'John Robert Seeley', *The English Historical Review*, 10 (1895), pp. 507–14.

Todhunter, M., 'Sir John Seeley', *Westminster Review*, 145 (1896), pp. 503–9.

Trevelyan, G. M., 'The Latest View of History', *Independent Review*, 1 (1903), pp. 395–414.

Ward, A. W., 'The Study of History at Cambridge', *Saturday Review*, 6 July 1872, pp. 9–10.

Ward, A. W., G. W. Prothero, and S. Leathes (eds), *The Cambridge Modern History*, vol. 1 *The Renaissance*, planned by Lord Acton (Cambridge: Cambridge University Press, 1902).

Secondary Sources

Altholz, J. L,. *Anatomy of a Controversy: The Debate over Essays and Reviews, 1860–1864* (Aldershot: Scolar Press, 1994).

Altholz, J. L., 'Lord Acton and the Plan of the *Cambridge Modern History*', *The Historical Journal*, 39 (1996), pp. 723–36.

Altick, R. D., *The English Common Reader: A Social History of the Mass Reading Public, 1800–1900*, 2nd edn (Columbus, OH: Ohio State University, 1998).

Amis, K., *Lucky Jim* (Harmondsworth: Penguin Books, 1961).

Bahners, P., '"A Place among the English Classics": Ranke's *History of the Popes* and its British Readers', in Stuchtey and Wende (ed.), *British and German Historiography 1750–1950*, pp. 123–57.

Bann, S., *The Clothing of Clio: A Study of the Representation of History in Nineteenth-Century Britain and France* (Cambridge: Cambridge University Press, 1984).

—, *The Inventions of History: Essays on the Representation of the Past* (Manchester and New York: Manchester University Press, 1990).

—, *Romanticism and the Rise of History* (New York: Twayne Publishers, 1995).

Barthes, R., 'The Discourse of History', trans. S. Bann, in *Comparative Criticism: A Yearbook*, ed. E. S. Shaffer, vol. 3 (Cambridge: Cambridge University Press, 1981), pp. 3–22.

Beatty, J., 'Replaying Life's Tape', *The Journal of Philosophy*, 103 (2006), pp. 319–36.

Bentley, M., *Modernizing England's Past: English Historiography in the Age of Modernism, 1870–1970* (Cambridge: Cambridge University Press, 2005).

—, 'Shape and Pattern in British Historical Writing, 1815–1945', in Macintyre, Maiguashca and Pók (eds), *The Oxford History of Historical Writing*, vol. 4.

—, 'British Historical Writing', in Scheider and Woolf (eds), *The Oxford History of Historical Writing*, vol. 5.

Bevington, M. M., *The Saturday Review, 1855–1868: Representative Educated Opinion in Victorian England* (New York: Columbia University Press, 1941).

Berger, S., 'The Invention of European National Traditions in European Romanticism', in Macintyre, Maiguashca and Pók (eds), *The Oxford History of Historical Writing*, vol. 4.

Blass, P. B. M., *Continuity and Anachronism: Parliamentary and Constitutional Development in Whig Historiography and in the Anti-Whig Reaction between 1890 and 1930* (The Hague: Nijhoff, 1978).

Brendon, P., *Hurrell Froude and the Oxford Movement* (London: Paul Elek, 1974).

Brooke, J., *Science and Religion: Some Historical Perspectives* (Cambridge: Cambridge University Press, 1991).

Browne, J., *Charles Darwin*, vol. 2: *The Power of Place* (London: Pimlico, 2003).

Brundage, A., *The People's Historian: John Richard Green and the Writing of History in Victorian England* (Westport, CT: Greenwood Press, 1994).

—, 'Green, John Richard (1837–1883)', in *Oxford Dictionary of National Biography* (Oxford: Oxford University Press, 2004).

Budge, G., 'Mackay, Robert William (1803–1882)' in *Oxford Dictionary of National Biography* (Oxford: Oxford University Press, 2004).

Butterfield, H., *Man on his Past: The Study of the History of Historical Scholarship* (Cambridge: University Press, 1969).

Burke, P., *History and Historians in the Twentieth Century* (Oxford: Oxford University Press, 2002).

Burrow, J. W., *A Liberal Descent: Victorian Historians and the English Past* (Cambridge: Cambridge University Press, 1981).

—, *A History of Histories: Epics, Chronicles, Romances and Inquiries from Herodotus and Thucydides to the Twentieth Century* (London: Penguin, 2007).

Campbell, J., 'Stubbs, Maitland, and Constitutional History', in Stuchtey and Wende (ed.), *British and German Historiography 1750–1950*, pp. 99–122.

—, 'Stubbs, William (1825–1901)', in *Oxford Dictionary of National Biography* (Oxford: Oxford University Press, 2004).

Cantwell, J. D., *The Public Record Office: 1838–1958* (London: HMSO, 1991).

Carignan, M., 'Analogical Reasoning in Victorian Historical Epistemology', *Journal of the History of Ideas*, 64 (2003), pp. 445–64.

Chadwick, O., *The Mind of the Oxford Movement* (California, CA: Stanford University Press, 1961).

—, *The Victorian Church*, vol. 1 (London: Adam & Charles Black, 1966).

—, 'Charles Kingsley at Cambridge', *The Historical Journal*, 18 (1975), pp. 303–25.

Cheng, E. K-M., 'Exceptional History? The Origins of Historiography in the United States', *History and Theory*, 47 (2008), pp. 200–28.

Collini, S., *Public Moralists: Political Thought and Intellectual Life in Britain 1850–1930* (Oxford: Clarendon Press, 1991).

—, D. Winch and J. Burrow, *That Noble Science of Politics: A Study in Nineteenth-Century Intellectual History* (Cambridge: Cambridge University Press, 1983).

Collingwood, R. G., *The Idea of History* (Oxford: Oxford University Press, 1970).

Confino, A., 'Narrative Form and Historical Sensation: On Saul Friedländer's *The Years of Extermination*', *History and Theory*, 48 (2009), pp. 199–219.

Cooter, R., *The Cultural Meaning of Popular Science: Phrenology and the Organization of Consent in Nineteenth-Century Britain* (Cambridge: Cambridge University Press, 1984).

Cooter, R. and S. Pumfrey, 'Separate Spheres and Public Places: Reflections on the History of Science Popularization and Science in Public Culture', *History of Science*, 32 (1994), pp. 237–67.

Copelman, D. M., *London's Women Teachers: Gender, Class and Feminism 1870–1930* (London: Routledge, 1996).

Corsi, P., *Science and Religion: Baden Powell and the Anglican Debate, 1800–1860* (Cambridge: Cambridge University Press, 1988).

Covert, J., *A Victorian Marriage: Mandell and Louise Creighton* (London: Hambledon/London: 2000).

Crowder, C. M. D., 'Creighton, Mandell (1843–1901)', *Oxford Dictionary of National Biography* (Oxford: Oxford University Press, 2004).

Culler, A. D., *The Victorian Mirror of History* (New Haven, Conn., Yale University Press, 1985).

Daston, L., 'Objectivity and the Escape from Perspective', *Social Studies of Science* 22 (1992), pp. 597–619.

Daston, L. and P. Galison, 'The Image of Objectivity', *Representations*, 40 (1992), pp. 81–128.

—, *Objectivity* (New York: Zone Books, 2007).

Desmond, A., *The Politics of Evolution: Morphology, Medicine, and Reform in Radical London* (Chicago, IL: University of Chicago Press, 1989).

—, *Huxley: From Devil's Disciple to Evolution's High Priest* (London: Penguin Books, 1997).

—, 'Redefining the X Axis: "Professionals", "Amateurs" and the Making of Mid-Victorian Biology—A Progress Report', *Journal of the History of Biology*, 34 (2001), pp. 3–50.

Dunn, W. H., *Froude & Carlyle: A Study of the Froude–Carlyle Controversy* (1930; Port Washington, NY: Kennikat Press, 1969).

—, *James Anthony Froude: A Biography*, 2 vols (Oxford: Clardendon Press, 1963).

Ellsworth, E. W., *Liberators of the Female Mind: The Shirreff Sisters, Educational Reform, and the Women's Movement* (Westport, CT: Greenwood Press, 1979).

Elton, G. R., *F. W. Maitland* (London: Weidenfeld and Nicolson, 1986).

—, *Return to Essentials: Some Reflections on the Present State of Historical Study* (Cambridge: Cambridge University Press, 2002).

Endersby, J., *Imperial Nature: Joseph Hooker and the Practices of Victorian Science* (Chicago, IL: University of Chicago Press, 2008).

Engel, A., 'Emerging Concepts of the Academic Profession at Oxford, 1800–1854', in L. Stone (ed.), *The University in Society*, vol. 1: *Oxford and Cambridge from the 14th to the Early 19th Century* (Princeton, NJ: Princeton University Press, 1974), pp. 305–51.

Evans, R. J., *In Defense of History* (New York: W. W. Norton and Company, 1999).

—, *Cosmopolitan Islanders: British Historians and the European Continent* (Cambridge: Cambridge University Press, 2009).

Forbes, D., *The Liberal Anglican Idea of History* (Cambridge: Cambridge University Press, 1953).

Frayn, M., *Copenhagen* (1998; New York: Anchor Books, 2000).

Fuchs, E., *Henry Thomas Buckle: Geschichtschreibung und Positivismus in England und Deutschland* (Leipzig: Leipziger Universitätsverlag, 1994).

—, 'Contemporary Alternatives to German Historicism in the Nineteenth Century', in Macintyre, Maiguashca and Pók (eds), *The Oxford History of Historical Writing*, vol. 4.

Gates, B. T., *Kindred Nature: Victorian and Edwardian Women Embrace the Living World* (Chicago, IL: University of Chicago Press, 1998).

Gates, B. T. and A. B. Shteir (ed.), *Natural Eloquence: Women Reinscribe Science* (Madison: University of Wisconsin Press, 1997).

Gieryn, T. F., 'Boundary-Work and the Demarcation of Science from Non-Science: Strains and Interests in Professional Ideologies of Scientists', *American Sociological Review*, 48 (1983), pp. 781–95.

—, *Cultural Boundaries of Science: Credibility on the Line* (Chicago, IL: University of Chicago Press, 1999).

Goldstein, D. S., 'J. B. Bury's Philosophy of History: A Reappraisal', *The American Historical Review*, 82 (1977), pp. 896–919.

—, 'The Professionalisation of History in Britain in the Late Nineteenth and Early Twentieth Centuries', *Storia della Storiographia*, 3 (1983), pp. 3–25.

—, 'The Origins and Early Years of the *English Historical Review*', *The English Historical Review*, 101 (1986), pp. 6–19.

—, 'History at Oxford and Cambridge: Professionalisation and the Influence of Ranke', in Iggers and Powell (eds), *Leopold von Rank and the Shaping of the Historical Discipline*, pp. 141–53.

—, 'Confronting Time: The Oxford School of History and the Non-Darwinian Revolution', *Storia della Storiografia*, 45 (2004), pp. 3–27.

Gould, S. J., *Wonderful Life: The Burgess Shale and the Nature of History* (New York: Norton, 1989).

Hacking, I. *The Taming of Chance* (Cambridge: Cambridge University Press, 1990).

Hanham, H. J., 'Editor's Introduction', in H. T. Buckle, *On Scotland and the Scotch Intellect*, ed. Hanham (Chicago, IL: University of Chicago Press, 1970).

Harvie, C., 'Bryce, James, Viscount Bryce (1838–1922)', *Oxford Dictionary of National Biography* (Oxford: Oxford University Press, 2004).

Hesketh, I., 'The Sobel Effect, Froude's Disease, and the Making of Un-Popular History in Mid-Victorian England', paper presented at the *History of Science Society Annual Conference*, November 2006, Vancouver, BC.

—, 'The Victorian Bible: *Ecce Homo* and the Manufacturing of a Literary Sensation', paper presented to the *Canadian Society of Church History Annual Conference*, June 2008, Vancouver, BC.

—, 'Diagnosing Froude's Disease: Boundary Work and the Discipline of History in Late-Victorian Britain', *History and Theory*, 47 (2008), pp. 373–97.

—, *Of Apes and Ancestors: Evolution, Christianity and the Oxford Debate* (Toronto: University of Toronto Press, 2009).

—, 'The Remains of the Freeman–Froude Controversy: The Religious Dimension', *Historical Papers 2010: Canadian Society of Church History* (forthcoming, 2010).

Heyck, T. W., *The Transformation of Intellectual Life in Victorian England* (London: St Martin's Press, 1982).

Hill, R., *Lord Acton* (New Haven, CT: Yale University Press, 2000).

Himmelfarb, G., *Lord Acton: A Study in Conscience and Politics* (Chicago, IL: University of Chicago Press, 1952).

Hinchliff, P., *God and History: Aspects of British Theology 1875–1914* (Oxford: Clarendon Press, 1992).

Howsam, L, 'Academic Discipline or Literary Genre? The Establishment of Boundaries in Historical Writing', *Victorian Literature and Culture* (2004), pp. 525–45.

—, 'Imperial Publishers and the Idea of Colonial History, 1870–1916', *History of Intellectual Culture* (2006), pp. 1–30.

—, *Past into Print: The Publishing of History in Britain 1850–1950* (London: The British Library; Toronto, University of Toronto Press, 2009).

Hull, D. L., *Darwin and His Critics: The Reception of Darwin's Theory of Evolution by the Scientific Community* (Cambridge, MA: Harvard University Press, 1973).

Iggers, G. G., 'Introduction', in L. von Ranke, *The Theory and Practice of History*, ed. Iggers and K. Von Moltke (Indianapolis, IN: Bobbs Merrill, 1973).

—, 'The Intellectual Foundations of Nineteenth-Century "Scientific History": The German Model', in Macintyre, Maiguashca and Pók (eds), *The Oxford History of Historical Writing*, vol. 4.

Iggers, G. G. and J. M. Powell (eds), *Leopold von Rank and the Shaping of the Historical Discipline* (Syracuse, NY: Syracuse University Press, 1990).

Jann, R., 'From Amateur to Professional: The Case of the Oxbridge Historians', *The Journal of British Studies*, 22 (1983), pp. 122–47.

—, *The Art and Science of Victorian History* (Columbus, OH: Ohio State University Press, 1985).

Janösi, F. E. De, 'The Correspondence between Lord Acton and Bishop Creighton', *Cambridge Historical Journal*, 6 (1939), pp. 307–13.

Jones, G. S., 'History: The Poverty of Empiricism', in R. Blackburn (ed.), *Ideology in Social Science: Readings in Critical Theory* (London: Fontana/Collins, 1972), pp. 96–115.

Kadish, A., 'Scholarly Exclusiveness and the Foundations of the *English Historical Review*', *Historical Research*, 61 (1988), pp. 183–97.

Kelley, D. R., *The Faces of History: Historical Inquiry from Herodotus to Herder* (New Haven, CT: Yale University Press, 1998).

—, *Fortunes of History: Historical Inquiry from Herder to Huizinga* (New Haven, CT: Yale University Press, 2003).

—, *Frontiers of History: Historical Inquiry in the Twentieth Century* (New Haven, CT: Yale University Press, 2006).

Kent, C., *Brains in Numbers: Elitism, Comtism, and Democracy in Mid-Victorian England* (Toronto: University of Toronto Press, 1978).

Kenyon, J., *The History Men: The Historical Profession in England since the Renaissance* (London: Weidenfeld and Nicolson, 1983).

Kitson Clark, G., 'The Origin of the *Cambridge Modern History*', *The Cambridge Historical Journal*, 8 (1945), pp. 57–64.

—, 'A Hundred Years of the Teaching of History at Cambridge, 1873–1973', *The Historical Journal*, 16 (1973), pp. 535–53.

Kochan, L., *Acton on History* (Port Washington, NY: Kennikat Press, 1954).

Krieger, L., *Ranke: The Meaning of History* (Chicago, IL: University of Chicago Press, 1977).

Krueger, C. L., 'Why She Lived at the PRO: Mary Anne Everett Green and the Profession of History', *Journal of British Studies*, 42 (2003), pp. 65–90.

Leeson, D., 'Cutting Through History: Hayden White, William S. Burroughs, and Surrealistic Battle Narratives', *Left History*, 10 (2004), pp. 13–43.

Lepenies, W., *Between Literature and Science: The Rise of Sociology* (Cambridge: Cambridge University Press, 1988).

Levine, G., *Dying to Know: Scientific Epistemology and Narrative in Victorian England* (Chicago, IL: University of Chicago Press, 2002).

Levine, P., *The Amateur and the Professional: Antiquarians, Historians and Archaeologists in Victorian England, 1838–1886* (Cambridge: Cambridge University Press, 1986).

—, *Feminist Lives in Victorian England: Private Roles and Public Commitment* (Oxford: Basil Blackwell, 1990).

Lightman, B. (ed.), *Victorian Science in Context* (Chicago, IL: University of Chicago Press, 1997).

—, 'Constructing Victorian Heavens: Agnes Clerk and the "New Astronomy"', in Gates and Shteir (eds), *Natural Eloquence*, pp. 43–75.

—, 'The Story of Nature: Victorian Popularizers and Scientific Narrative', *Victorian Review*, 25:2 (1999), 1–29.

—, *Victorian Popularizers of Science: Designing Nature for New Audiences* (Chicago, IL: University of Chicago Press, 2007).

—, *Evolutionary Naturalism in Victorian Britain: The 'Darwinians' and Their Critics* (Berlington, VT: Ashgate, 2009).

Looser, D., *British Women Writers and the Writing of History 1670–1820* (Baltimore, MD: Johns Hopkins University Press, 2000).

Lowenthal, D., *The Past is a Foreign Country* (Cambridge: Cambridge University Press, 1985).

MacDougall, H. A., *Racial Myth in English History: Trojans, Teutons, and Anglo-Saxons* (Montreal: Harvest House, 1982).

Macintyre, S., J. Maiguashca and A. Pók (eds), *The Oxford History of Historical Writing*, vol. 4: *1800–1945* (Oxford: Oxford University Press, forthcoming, 2011).

Markus, J. *J. Anthony Froude: The Last Undiscovered Great Historian* (New York: Scribner, 2005).

McCartney, D., *W. E. H. Lecky: Historian and Politician 1838–1903* (Dublin: The Lilliput Press, 1994).

McCaw, N., *George Eliot and Victorian Historiography: Imagining the National Past* (London: Macmillan Press, 2000).

McDowell, R. B., *Alice Stopford Green: A Passionate Historian* (Dublin: Allen Figgis and Company, 1967).

McLachlan, J. O., 'The Origin and Development of the Cambridge Historical Tripos', *Cambridge Historical Journal*, 9 (1947), pp. 78–105.

McNeill, R., 'Froude and Freeman', *Monthly Review*, 22 (1906), pp. 79–91.

Meadows, J., *The Victorian Scientist: The Growth of a Profession* (London: The British Library, 2004).

Miller, D. P., 'The "Sobel Effect": The Amazing Tale of How Multitudes of Popular Writers Pinched All the Best Stories in the History of Science and Became Rich and Famous while Historians Languished in Accustomed Poverty and Obscurity, and How this Transformed the World', *Metascience* (2002), pp. 185–200.

Mitchell, R., *Picturing the Past: English History in Text and Image 1830–1870* (Oxford: Clarendon Press, 2000).

Murphy, J. M., *Positivism in England: The Reception of Comte's Doctrines, 1840–1870* (New York: Columbia University, 1968).

Nagel, T., *The View from Nowhere* (Oxford: Oxford University Press, 1986).

Nockles, P. B., *The Oxford Movement in Context: Anglican High Churchmanship, 1760–1857* (Cambridge: Cambridge University Press, 1994).

Norton, B., *Freeman's Life: Highlights, Chronology, Letters and Works* (Farnborough: Norton, 1993).

Novick, P., *That Noble Dream: The 'Objectivity Question' and the American Historical Profession* (Cambridge: Cambridge University Press, 1988).

Owen, D. M., 'The Chichele Professorship of Modern History, 1862', *Bulletin of the Institute of Historical Research*, 34 (1961), pp. 217–20.

Parker, C., 'English Historians and the Opposition to Positivism', *History and Theory*, 22 (1983), pp. 120–45.

—, *The English Historical Tradition since 1850* (Edinburgh: John Donald, 1990).

Paul, H., *The Life of Froude* (London: Sir Isaac Pitman & Sons, 1905).

Pérez-Ramos, A., *Francis Bacon's Idea of Science and the Maker's Knowledge Tradition* (Oxford: Clarendon Press, 1988).

Peterson, M. J., *Family, Love and Work in the Lives of Victorian Gentlewomen* (Indianapolis, IN: Indiana University Press, 1989).

Philips, M. S., *Society and Sentiment: Genres of Historical Writing in Britain, 1740–1820* (Princeton, NJ: Princeton University Press, 2000).

Pocock, J. G. A., *The Ancient Constitution and the Feudal Law: A Study of English Historical Thought in the Seventeenth Century* (Cambridge: Cambridge University Press, 1957).

Porter, T., *The Rise of Statistical Thinking, 1820–1900* (Princeton, NJ: Princeton University Press, 1986).

—, *Karl Pearson: The Scientific Life in a Statistical Age* (Princeton, NJ: Princeton University Press, 2004).

—, 'Buckle, Henry Thomas', in B. Lightman (ed.), *The Dictionary of Nineteenth-Century British Scientists*, vol. 1 (Bristol: Thoemmes Continuum, 2004).

—, 'The Objective Self', *Victorian Studies*, 50 (2008), pp. 641–7.

Read, G., 'Reading Obituaries: Death, Masculinity and Republicanism in Interwar France, 1919–1940', paper presented at the New Frontiers in Graduate History Conference, February 2004, York University, Toronto, Canada.

—, '*Des Hommes et des citoyens*: Paternalism and Citizenship on the Republican Right in Interwar France, 1919–1940', *Historical Reflections/Reflexions Historiques*, 34 (2008), pp. 88–111.

Rich, P. B., *Race and Empire in British Politics*, 2nd edn (1986; Cambridge: Cambridge University Press, 1990).

Rigney, A., *Imperfect Histories: The Elusive Past and the Legacy of Romantic Historicism* (Ithaca, NY: Cornell University Press, 2001).

Ross, D., 'On the Misunderstanding of Ranke and the Origins of the Historical Profession in America', in G. G. Iggers and J. M. Powell (eds), *Leopold von Ranke and the Shaping of the Historical Discipline* (Syracuse, NY: Syracuse University Press, 1990), pp. 154–69.

Rossiter, M. W., '"Women's Work" in Science, 1880–1910', *Isis*, 71 (1980), pp. 381–98.

Rothblatt, S., *The Revolution of the Dons: Cambridge and Society in Victorian England* (London: Faber and Faber, 1968).

Rudwick, M. J. S., *The Great Devonian Controversy: The Shaping of Scientific Knowledge among Gentlemanly Specialists* (Chicago, IL: University of Chicago Press, 1985).

Ruse, Michael, 'Darwin's Debt to Philosophy: An Examination of the Influence of the Philosophical Ideas of John F. W. Herschel and William Whewell on the Development of Charles Darwin's Theory of Evolution', *Studies in the History and Philosophy of Science*, 6 (1975), pp. 159–81.

—, *The Darwinian Revolution: Science Red in Tooth and Claw*, 2nd edn (1979; Chicago, IL: University of Chicago Press, 1999).

Sastri, K. A. N. and H. S. Ramanna, *Historical Method in Relation to Indian History* (Madras: S. Viswanathan, 1956).

Schneider, A. and D. Woolf (eds), *The Oxford History of Historical Writing*, vol. 5: *Historical Writing since 1945* (Oxford: Oxford University Press, forthcoming, 2011).

Schneider, A. and S. Tanaka, 'The Transformation of History in China and Japan', in Macintyre, Maiguashca, and Pók (eds), *The Oxford History of Historical Writing*, vol. 4.

Scott, J. W., *Gender and the Politics of History* (New York: Columbia University Press, 1988).

Secord, J. A., *Victorian Sensation: The Extraordinary Publication, Reception, and Secret Authorship of* Vestiges of the Natural History of Creation (Chicago, IL: University of Chicago Press, 2000).

Semmel, B., 'H. T. Buckle: The Liberal Faith and the Science of History', *British Journal of Sociology*, 27 (1976), pp. 370–86.

Shapin, S., *A Social History of Truth: Civility and Science in Seventeenth-Century England* (Chicago, IL: University of Chicago Press, 1994).

Shapin, S. and S. Schaffer, *Leviathan and the Air-Pump: Hobbes, Boyle, and the Experimental Life* (Princeton, NJ: Princeton University Press, 1985).

Simon, W. M., *European Positivism in the Nineteenth Century: An Essay in Intellectual History* (Port Washington, NY: Kennikat Press, 1972).

Slee, P. R. H., *Learning and a Liberal Education: The Study of Modern History in the Universities of Oxford, Cambridge, and Manchester, 1840–1914* (Manchester: Manchester University Press, 1986).

Small, H., 'Chances Are: Henry Buckle, Thomas Hardy, and the Individual at Risk', in Small and T. Tate (eds), *Literature, Science, Psychoanalysis, 1830–1970* (Oxford: Oxford University Press, 2005), pp. 64–85.

Smith, B. G., *The Gender of History: Men, Women, and Historical Practice* (Cambridge, MA: Harvard University Press, 1998).

Smith, C., *The Science of Energy: A Cultural History of Energy Physics in Victorian Britain* (Chicago, IL: University of Chicago Press, 1998).

Smith, J., *Fact and Feeling: Baconian Science and the Nineteenth-Century Literary Imagination* (Madison, WI: University of Wisconsin Press, 1994).

Snyder, L. J., *Reforming Philosophy: A Victorian Debate on Science and Society* (Chicago, IL: University of Chicago Press, 2006).

Soffer, R. N., *Discipline and Power: The University, History, and the Making of an English Elite, 1870–1930* (Stanford, CA: Stanford University Press, 1994).

—, *History, Historians and Conservatism in Britain and America: The Great War to Thatcher and Reagan* (Oxford: Oxford University Press, 2009).

Southgate, B., *History Meets Fiction* (Harlow: Pearson, 2009).

Spahr, E. and R. J. Swenson, *Methods and Status of Scientific Research with Particular Applications to the Social Sciences* (New York: Harper & Brothers Publishing, 1930).

St. Aubyn, G., *A Victorian Eminence: The Life and Works of Henry Thomas Buckle* (London: Barrie, 1958).

Stephens, H. M., *Counsel upon the Reading of Books* (1900; Port Washington, NY: Kennikat Press, 1968).

Stross, W. A., 'Magazines of Mortality: A Cultural History of the Obituary in Eighteenth-Century London', PhD diss., University of Toronto, 2004.

Stuchtey, B. and P. Wende (eds), *British and German Historiography 1750–1950: Traditions, Perceptions, and Transfers* (Oxford: Oxford University Press, 2000).

Tollebeek, J., 'Seeing the Past with the Mind's Eye: The Consecration of the Romantic Historian', *Clio*, 29:2 (2000), pp. 167–91.

Topham, J., 'Scientific Publishing and the Reading of Science in Nineteenth-Century Britain: A Historiographical Survey and Guide to Sources', *Studies in the History and Philosophy of Science*, 31:4 (2000), pp. 559–612.

Tosh, J., *Manliness and Masculinities in Nineteenth-Century Britain* (Harlow: Pearson, 2005).

Tulloch, H., 'Lord Acton and German Historiography', in Stuchtey and Wende (eds), *British and German Historiography 1750–1950*, pp. 159–72.

Turner, F. M., *Contesting Cultural Authority: Essays in Victorian Intellectual Life* (Cambridge: Cambridge University Press, 1993).

Vance, N., 'Kingsley, Charles (1819–1875)', *Oxford Dictionary of National Biography* (Oxford: University Press, 2004).

Vernon, J., 'Narrating the Constitution: The Discourse of "the Real" and the Fantasies of Nineteenth-Century Constitutional History', in Vernon (ed.), *Re-Reading the Constitution: New Narratives in the Political History of England's Long Nineteenth Century* (Cambridge: Cambridge University Press, 1996), pp. 204–29.

Von Arx, J. P., *Progress and Pessimism: Religion, Politics, and History in Late Nineteenth Century Britain* (Cambridge, MA: Harvard University Press, 1985).

Watson, G., *Lord Acton's History of Liberty* (Aldershot: Scolar Press 1994).

White, H., 'The Burden of History', *History and Theory*, 5 (1866), pp. 111–34.

—, *Tropics of Discourse: Essays in Cultural Criticism* (Baltimore, MD: Johns Hopkins University Press, 1985).

—, *The Content of the Form: Narrative Discourse and Historical Representation* (Baltimore, MD: Johns Hopkins University Press, 1987).

White, P., *Thomas Huxley: Making the 'Man of Science'* (Cambridge: Cambridge University Press, 2003).

Whitney, F. L., *The Elements of Research* (New York: Prentice-Hall, 1942).

Williams, N. J., 'Stubbs's Appointment as Regius Professor, 1866', *Bulletin of the Institution of Historical Research*, 33 (1960), pp. 121–5.

Winter, A., 'The Construction of Orthodoxies and Heterodoxies in the Early Victorian Life Sciences', in Lightman (ed.), *Victorian Science in Context*, pp. 24–50.

—, *Mesmerized: Powers of Mind in Victorian Britain* (Chicago, IL: University of Chicago Press, 1998).

Wright, T. R., *The Religion of Humanity: The Impact of Comtean Positivism on Victorian Britain* (Cambridge: Cambridge University Press, 1986).

Woolf, D., 'A Feminine Past? Gender, Genre, and Historical Knowledge in England, 1500–1800', *American Historical Review*, 102 (1997), pp. 645–79.

—, *A Global History of History* (Cambridge: Cambridge University Press, forthcoming, 2011).

Wormell, D., *Sir John Seeley and the Uses of History* (Cambridge: Cambridge University Press, 1980).

Yeo, R., 'Science and Intellectual Authority in Mid-Nineteenth-Century Britain: Robert Chambers and "Vestiges of the Natural History of Creation"', *Victorian Studies*, 28 (1984), pp. 5–31.

—, 'An Idol of the Market-Place: Baconianism in Nineteenth-Century Britain', *History of Science*, 33 (1985), pp. 251–98.

—, *Defining Science: William Whewell, Natural Knowledge and Public Debate in Early Victorian Britain* (Cambridge: Cambridge University Press, 1993).

INDEX

Academy, 113, 140, 144–5, 150–61, 160
Acton, Lord, 21, 35, 37, 49, 95, 132, 163
  as co-editor of *Rambler*, 37, 39, 148
  as editor of the *Home and Foreign Review*, 99
  as Regius Professor of Modern History at Cambridge, 96, 110, 119, 155, 161–2
  biographical sketch of, 37–8
  Bryce and, 130, 152
  Creighton and, 101–4, 106–7, 112
  death of, 133, 147–8
  debate with Creighton, 8–9, 101, 108–10, 127
  Döllinger and, 37–8, 96, 149
  EHR and, 100–3, 106–7, 110, 130, 149
  inaugural lecture of, 97–9, 150
  literary paralysis and, 10, 149–51
  Macaulay and, 97
  on Buckle, 37, 39–44, 53
  on objectivity, 98–9, 108–10, 111–13, 149, 155
  on morality and history, 8–9, 96, 98–9, 101, 108–11, 113, 149
  on science and Catholicism, 39, 44
  on the science of history, 3–5, 43–4, 96–9, 147, 149
  Poole and, 147–51
  posthumous opinions of, 147–51
  Stubbs and, 45, 98, 111, 142
  style and art of, 150–1
  Ranke and, 38, 97, 106, 149
  works of,
    Cambridge Modern History, 7, 110–14, 148–9, 161
    'German Schools of History', 106–7
Acton, Richard, 37

Amis, Kingsley,
  Lucky Jim, 164
Anglican faith *see* Church of England
Annales, school of, 163
Antiquaries, Society of, 133
Athenaeum, 77, 135, 143–6, 148–50
Archaeological Journal, 135
Arnold, Matthew, 160–1
Arnold, Thomas, 92
art of history, 4, 6–7, 9, 86, 152–3, 163–4
  in Acton, 150–1
  in Buckle, 38
  in Froude, 4, 65, 71–2, 157, 161
  in Stubbs, 144, 146
  critique of, 73–4, 80, 144
archives, 106, 163
  expansion of, 3, 35, 113
  Kingsley and, 61
  Freeman and, 118
  Froude and, 6, 64, 68, 70–1, 87, 156–8
  Stubbs and, 4, 45, 48, 55, 142
audience, concern with, 8, 73–4, 99–100, 106, 115–16, 159
  Buckle and, 31–2
  Creighton and, 130–2
  EHR and, 130–2
  Freeman and, 5–6, 8, 81–2, 86–7, 91, 95, 117–19, 124
  Green and, 121–3
  Seeley and, 6, 76, 78–82, 95, 128–30

Bacon, Francis, 3, 22, 36, 48
Baconianism, 3–4, 23, 36, 38, 52
  see also induction
Bann, Stephen, 49
Barlow, John, 13
Barthes, Roland, 112

– 219 –

Beattie, William, 15
Becket, Thomas, 87–9, 91
belles lettres, 4, 80, 134
Bentham, Jeremy, 24–5
biblical criticism, 16, 19, 57, 66, 75
Bible, 66, 75
   Old Testament of, 66
   see also biblical criticism; Christianity; God
Bluntschli, Johann Kaspar, 149
boundary work, 5–6, 8, 9, 37, 73, 85, 119, 164
   definition of, 86
   Buckle as a casualty of, 5, 37, 73, 153
   Froude as a casualty of, 6, 85–8, 93, 95, 153, 161
   Kingsley as a casualty of, 6, 74, 95
Boyle, Robert, 3
Bridgewater Treatises, 16, 27
Brewster, David, 33
British Quarterly Review, 61
Broad Church, 57–8
Böckh (Boeckh), Philipp August, 149
Brougham, Henry, 22
Browning, Robert, 161
Bryce, James, 95, 100, 115
   Acton and, 130, 152
   biographical sketch of, 104
   Creighton and, 101, 103–7
   EHR and, 104–6, 130–1, 138, 138, 152
   Freeman and, 90–2, 104, 117, 124–7, 134–8, 141, 151
Buckle, Henry Thomas, 4, 10, 38, 50, 53–4
   Acton and, 37, 39–44
   as a casualty of boundary work, 5, 37, 73, 153
   as celebrity, 14, 39
   as English Comte/positivist, 3, 19, 36, 39, 41–2, 44, 164
   biographical sketch of, 15–16
   concern with audience, 30–1
   criticism of Comte, 1, 20, 24
   death of, 32–3, 153
   debt to Comte, 1, 17, 19
   Droysen and, 43–4
   Freeman and, 53
   Froude and, 70, 72
   Illness of, 15, 31–2

   influence and critique of Mill, 33–5
   Kingsley and, 58–9, 61
   Mackay and, 33
   on deduction, 27–8, 30
   on great men, 32–3
   on imagination and history, 2, 25–6, 28–9, 36
   on induction/Baconianism, 23, 27, 27, 30, 36
   on morality, 14–15, 21, 44, 51, 59
   on Newton's discovery of gravity, 28–9
   on statistics, 20–2, 79
   on the discipline of history, 19, 35–6, 42
   on the role of poetry and science in English history, 25–6
   on the role of the environment, 18
   on the role of women and science, 26–7
   posthumous opinions of, 33
   rejection of race concept, 51
   Royal Institution lecture ('The Influence of Women on the Progress of Knowledge'), 13–14, 16–19
   Seeley and, 53, 78–9
   Stubbs and, 33, 35–6, 47
   Whewell and, 16, 25, 29
   see also *History of Civilization in England*
Buckle, Thomas Henry, 16
Burd, L. A., 150
Burke, Edmund, 24, 41, 50
Burrows, Montagu,
   as Chichele Professor of Modern History, 82–5
   Pass and Class, 83–5
Bury, J. B., 161–3

Catholicism, 2, 37–8, 45, 57, 68, 75, 96–7, 99
   Counter-Reformation, 8, 108
   science and, 39, 44
   Tractarian Movement and, 56, 65
   see also Acton, Lord; Newman, John Henry
Cambridge Modern History, 7, 115, 148–9, 161
   Creighton's 'Introductory Note' of, 112–14
   method of, 111–12
   plan of, 110–11, 114
   see also Acton, Lord

Cambridge Review, 149
Cambridge University, 37, 99, 102, 119, 138, 140
   Dixie Professorship of History at, 100, 127
   Magdalene College of, 56
   Regius Professors of Modern History at, 38, 55–6, 74, 76–7, 95–6, 110, 128, 154, 161–3
   reform of, 96–7
   study of history at, 62, 74, 78, 96–7, 163
Cambridge University Press, 110
Carignan, Michael, 36
Carlingford, Lord, 119
Carlyle, Thomas, 3, 6, 12, 58, 65, 67, 73, 82, 164
   Froude and, 67–70, 72, 156
Carnarvon, Lord, 46
Chambers, Robert,
   Vestiges of the Natural History of Creation, 5, 30, 33, 86
Chapman, John, 33
Chartism, 57
Christianity, 42, 55, 65, 120, 136
   Christ's divinity and, 66, 75
   crisis of faith in, 67
   science and, 17, 48–9, 57, 75
   science of history and, 37, 48–9
   morality and, 9, 44, 108, 110
   see also Jesus Christ; Catholicism; God; Protestantism; Providence; muscular Christianity
Christian Observer, 65
Christien, John, 75
Chronicles and Memorials of Richard I *see* Rolls Series
Church of England, 19, 56–7, 64–7, 75, 88, 92, 100
   declining role at ancient universities, 97
   see also Protestantism; Thirty-Nine Articles; Tractarian Movement
Clarendon Press *see* Oxford University Press
Clarendon Press Series, 117
Clarkson, Thomas, 22
Coleridge, Samuel Taylor, 24–5
Collini, Stefan, 115
Combe, George, 86–7

Comte, August, 42
   Cours de philosophie positive, 1
   on theology, 18
   Religion of Humanity and, 1, 17, 19–20, 24, 31
   stages of human development, 18
   Victorian intellectuals and, 17
Condorcet, 19
Contemporary Review, 87
contingency, 10–11
Cook, John Douglas, 47
covering law, theory of, 163
Creighton, Louise, 127
Creighton, Mandell, 95, 115, 128, 163
   Acton and, 101–4, 106–7, 112
   as editor of the *EHR*, 100, 130–2
   Bryce and, 101, 103–7
   biographical sketch of, 100
   Cambridge Modern History and, 112–13
   concern with audience, 131–2
   death of, 133, 147
   debate with Acton, 8, 108–10
   Gladstone and, 131
   Longman and, 127, 130–2
   on the science of history, 113–14, 147
   works of,
      Age of Elizabeth, 127
      A History of the Papacy, 100, 107
      Primer of Roman History, 127
      The Shilling History of England, 127
      Tudors and the Reformation, 127

Dalberg, Marie, 37
Darwin, Charles, 29, 50
   Green and, 120
   on Buckle, 30
   on Kingsley, 57
   On the Origin of Species, 14, 17, 20
Daston, Lorraine, 98
deduction, method of, 30
   gender of, 27–8
   role in scientific discovery, 28–9
determinism, 14
Dickens, Charles, 2, 7, 36
Dilthey, Wilhelm, 149
disinterest, theory of, *see* objectivity
Disraeli, Benjamin, 46

Döllinger, Johann Joseph Ignaz von, 37–8, 96, 149
Droysen, Johann Gustav, 43, 106, 149

Eclectic Review, 63
Edinburgh Review, 53, 68, 87, 148, 150–61
Edward III, 163
English Historical Review (*EHR*), 100–3, 143, 149, 161
  as a venue for scientific history, 7, 105–6, 115, 131–2, 153
  Acton and, 7, 107, 109–10, 112
  audience for, 8, 130–2
  Creighton and, 130–2
  Poole and, 141, 147–8, 154
  'Prefatory Note' of, 104–6, 138, 151
  obituaries and, 9, 133–4, 138, 141–2, 147, 151, 157
English Illustrated Magazine, 157
Edinburgh University, 86
Education Act of 1870, 125
Egerton, Francis Henry (Earl of Bridgewater), 16
Eliot, George, 17, 33
Elton, G. R., 77
Essays and Reviews, 16–17, 47, 55, 58, 66
Evans, John, 133
evolution, 5, 17, 50, 57
evolutionary naturalists, 17
Examiner, 31

Fairbairn, A. M., 111
Firth, C. H., 154, 156
Fisher, H. A. L.
  on Froude, 158–60
  on Seeley, 139–40
Fortnightly Review, 158
Foster, John, 31
Freeman, Edward A., 64, 73, 78, 100–2, 109, 115, 147, 153, 163–4
  architecture and, 136–7
  as disciplinarian, 6, 74, 80, 135
  as professional historian, 83–4, 93, 134
  as Regius Professor of Modern History at Oxford, 85, 92–3, 154–5, 157–8
  as unpopular professor, 85, 158
  as whig historian, 9
  biographical sketch of, 53
  Bryce and, 90–2, 104, 117, 124–7, 134–8, 141, 151–2
  Chichele Professorship of Modern History at Oxford and, 82–5
  concern with audience, 5–6, 8, 81–2, 87, 91, 95, 117–19, 124, 128
  death of, 133
  ecclesiology and, 52, 81
  Froude's criticism of, 89–90
  geology and, 136–7
  Green and, 92, 118–27
  Macmillan and, 116–19, 124, 126
  on accuracy, 52, 81, 87–8, 92, 117–18, 134–7
  on Buckle, 53
  on Carlyle, 82
  on continuity in English history, 52–3
  on Froude, 6, 82, 85–93, 134, 154
  on Kingsley, 58, 62–3, 117, 134
  on morality and history, 73, 92, 134
  on scientific racism, 52–3
  on the science of history, 3, 81–2, 95, 98, 147
  philological method of, 51–2, 137
  posthumous opinions of, 134–8
  role of imagination in, 137
  Saturday Review and, 47, 134, 137
  Seeley and, 138
  Stubbs and, 35–6, 46, 48, 52–3, 90–2, 120–2, 142
  style of, 135, 138
  works,
    'Historical Course for Schools', 124–7
    History of Federal Government, 116, 118
    History of the Norman Conquest, 52–3, 117–21, 137, 149
    Old English History for Children, 116–19, 121, 129
free-will, 18, 21, 40, 44, 53
Frazer's Magazine, 69, 123
Froude, James Anthony, 46, 48, 55, 73, 83–4, 147
  archival research of, 6, 64, 68, 70–81, 87, 156–8
  as casualty of boundary work, 6, 85–8, 93, 95, 153–4, 161
  as controversial novelist, 64–7, 70

as excluded from the *EHR*, 101, 106,
    153, 157
as Regius Professor of Modern History at
    Oxford, 93, 154–5, 157–8, 163
as whig historian, 9
biographical sketch of, 65–8
Carlyle and, 67–70, 72, 156
Freeman's attacks on, 6, 82, 85–93, 134
inaugural lecture of, 155–6
Kingsley and, 67–70, 72
Newman and, 65, 67–8
on Buckle, 4, 70
on Elizabethan England, 68
on Freeman, 89–70
on Henry VIII, 68–9, 87, 89
on history as art/drama, 4, 7, 65, 71–2,
    86, 139, 144, 154–7, 159–61, 164
on morality and history, 71, 156, 160
on the science of history, 4, 7, 70–2,
    155–6, 159–60
posthumous opinions of, 158–61
role of imagination in, 68–9, 72, 89, 160
Smith and, 68–71, 87
style of, 68, 70, 86, 154, 158
works of,
    Divorce of Catherine of Aragon, 153
    History of England, 64, 68–9, 87, 89,
        91, 156–7
    Life and Letters of Erasmus, 158
    'Life and Times of Thomas Becket',
        87–9, 91, 154
    The Nemesis of Faith, 66–7, 88
    Shadows of the Clouds, 66
    Spanish Story of the Armada, 154
    see also 'Froude's disease'
Froude, Robert Hurrell, 64–5
Froude, Richard Hurrell, 64, 67, 91
    *Remains of the Reverend R. Hurrell
        Froude*, 65, 89
    see also Tractarian Movement
'Froude's disease', 6, 90, 93, 161
Fustel de Coulanges, Numa Denis, 98

Galison, Peter, 98
Gardiner, Betha Meriton Cordery, 127
Gardiner, Samuel, 127, 147, 154, 163
    History of England, 82
Gasquet, Francis Aidan, 111

Gibbon, Edward, 160
Gieryn, Thomas, 37, 86
Gladstone, W. E., 45, 83, 115
    Seeley and, 76–7
    Creighton and, 131
Glasgow University, 104
God, 14, 16, 40, 43, 49, 53, 57, 61, 66
    see also Jesus Christ; Christianity
Goethe, Johann Wolfgang von, 71
Gooch, G. P., 148–51
Gould, Stephen Jay, 10
Gradgrind, Thomas see Dickens, Charles
gravity, law of, 2, 28–9
    see also Newton, Isaac
great man, theory of, 22–3, 68–9, 106,
    156–7
    see also Carlyle, Thomas
Green, Alice Stopford, 127–8
Green, J. R., 147
    biographical sketch of, 119–20
    Bryce and, 104
    Freeman and, 92, 118–27
    as Darwinian, 120
    concern with audience, 121–3, 128
    historical reviews and, 100–11
    Kingsley and, 77
    Macmillan and, 121–3, 126–7
    the *Saturday Review* and, 47, 120
    Seeley and, 77, 130
    Stubbs and, 48, 120, 122, 142
    works of
        Short History of the English People,
            121–4, 127, 130
Greville, Charles
    Greville Memoirs, 131
Grey, Lord, 22
Grey, Maria, 1, 32
Grimm, Jacob, 51
Grote, Harriet, 32
Grove, George, 122
Guardian, 137

Hacking, Ian, 20
Hallam, Henry, 48, 142
    death of, 147
Harold II, 118, 142
Harrison, J. B., 111
Harvey, William, 26

Häser, Heinrich, 42
Hecker, Justus Friedrich Karl, 42
Hegel, G. W. F., 106
Henry VIII, 56, 68, 87
hero-worship, theory of, *see* great man, theory of
Hermann, Franz, 149
Heyck, T. W., 37
History of Civilization in England, 1, 13–14
    Acton's criticism of, 37, 39–44
    Droysen's criticism of, 43–4
    Buckle's failure to complete, 32
    England in, 19
    general argument, 16–19
    plan of, 15
    possible posthumous volume of, 24
    on great men versus the masses, 12–23, 40–51
    role of poetry in, 25–6
    role of statistics in, 20–1, 40–1
    rejection of race concept in, 51
    Stubbs's criticism of, 35–6, 47
    Vestiges of the Natural History of Creation and, 30, 33
    Whewell in, 35
    writing style of, 30–1, 39
    See also Buckle, Henry Thomas
Historische Zeitschrift, 107
Holyoake, George, 30
Holy Roman Empire, 104
Home and Foreign Review, 99
Hooker, Joseph, 30
Hope, A. J. B. Beresford, 47
Howsam, Leslie, 8, 119, 126
Hume, Martin, 154
Hunt, Leigh, 15
Huth, Alfred Henry, 13, 19
Huxley, Thomas Henry, 5, 8, 18
    Wilberforce and, 120

Iggers, Georg G., 38
imagination, role of, 9, 162
    in Buckle, 2, 25–6, 28–9, 36
    in Freeman, 137
    in Froude, 68–9, 72, 89, 160
    in Macaulay, 79
    in Romantic history, 4–5
    in Seeley, 140

induction, 23, 30, 38, 44
    as method of history, 4–7, 9–10, 47, 50, 52–3, 55, 60, 64, 70, 72–3, 78, 95, 99, 108, 113, 115, 143, 151, 155
    Christianity and, 44, 49
    gender of, 27–9
    Whewell and, 25
    see also Baconianism

Jessop, Augustus, 153–4, 158
Jesus Christ, 57
    history of, 74–5
    see also Seeley, J. R.
Jowett, Benjamin, 66

Keble, John, 56, 65, 155
Kemble, John, 48
    The Saxons in England, 51
Kepler, Johannes,
Kingsley, Charles, 116, 119
    as casualty of boundary work, 6, 73, 85, 95
    as controversial novelist, 55–6, 62
    as member of the 'Broad Church', 57–8, 89
    as 'muscular Christian', 58, 69
    as popular lecturer, 62, 64
    as Regius Professor of Modern History at Cambridge, 55–6, 58, 61, 74, 76–8, 85, 154
    critique of Buckle, 4, 58–9
    Freeman's attacks of, 58, 62–3, 74, 117, 134
    Froude and, 67–70, 72
    historical method of, 4, 58, 60–1
    inaugural lecture of, 58–62
    on the Tractarians, 56
    Seeley and, 74–6
    Stubbs and, 58, 144
    works of,
        Alton Locke, 56–7
        Hypatia: New Foes with an Old Face, 56
        Roman and the Teuton, 62–4
        Two Years Ago, 58
        The Water-Babies, 63
Kingsley, Mary, 75
Kingsley, Rose, 75
Kintore, Lord, 15

Knight, Charles,
　Popular History of England, 122
Knowles, James, 96
Krieger, Leonard, 38

Lambeth Palace, 120
Lang, Andrew, 153
Langlois, Charles Victor,
　Introduction to the Study of History, 93
Lassen, Christian, 42
Leveson-Gower, George (Earl of Granville), 37
Lewes, George Henry, 17, 32
Liebermann, Felix, 111
Lily, W. S., 96–7
linguistic turn, 163
Literary Gazette, 58, 61
Literary Opinion, 135, 137
London Review, 62, 76
Longman, Charles, 101
　Creighton and, 127, 130–2
Louis XIV, 41
Luther, Martin, 22, 60
Lyell, Charles, 17
Lyttleton, Lord, 15

Macaulay, Thomas Babington, 3, 6, 23, 48, 73, 144, 163
　Acton and, 97
　as whig historian, 9
　death of, 147
　History of England, 123
　Seeley and, 79–80, 85
Mackay, Robert William, 33
Macmillan, Alexander,
　Green and, 100, 121–3, 126–7
　Freeman and, 116–19, 124, 126
　Seeley and, 75–6, 128–30
Machiavelli, Niccolò, 150
Macmillan's Magazine, 78, 128
Magna Carta, 142
Maitland, F. W., 133, 138
　on Acton, 149–51
　on Stubbs, 141–7, 151
　The History of English Law, 142
Malthus, Thomas, 24
Manchester, University of, 154
Mansel, Henry Longueville, 46

Markham, C., 136–7
Markus, Julia, 65
metaphysics, 18
men,
　and science, 27–8
　'nature' of, 28
Michelet, Jules, 99
Mill, John Stuart, 15, 17, 19, 33
　Buckle and, 24–5
　education of, 24
　race concept and, 51
　works of,
　　On Liberty, 33
　　System of Logic, 33
Mommsen, Theodor, 85, 97, 106
Montesquieu, 18–19
Montfort, Simon de, 127
morality and history, treatment of, 3, 7–9, 95
　by Acton, 8–9, 96, 98–9, 101, 108–11, 113, 149
　by Buckle, 14–15, 21, 44, 51, 59
　by Freeman, 73, 92, 134
　by Froude, 71, 156, 160
　Müller, Max, 51
　on Kingsley, 63–4
Munich University, 37
muscular Christianity, 58, 69
　see also Kingsley, Charles; Christianity

Nagel, Thomas, 99
narrative form of history, 3, 5–6, 9–10, 41, 49–50, 68, 79–82, 88, 96–8, 108, 110–11, 113–14, 123, 129, 132, 151
natural theology, 5
　and induction, 48–9
　see also Paley, William
Niebuhr, Reinhold, 85, 106, 162
Nineteenth Century, 89, 91, 96
Newman, John Henry, 39, 65, 67, 146, 161
　Tract 90, 56
　see also Tractarian Movement
Newton, Isaac, 26
　discovery of gravity and, 2, 28–9
Norman Conquest, 52, 117, 119, 143
　see also Freeman, Edward A.
North British physicists, 5

objectivity, 4, 6, 70, 83, 151, 155, 158–9, 164
   mechanical form of, 7, 98–9
   Old Testament, 66
   see also, Bible
Oscott College, 37
Oxford Chronicle, 119
Oxford Movement *see* Tractarian Movement
Oxford University, 37, 64–5, 102
   Balliol College of, 117
   Chichele Professorship of Modern History at, 46, 82–5
   Christ Church of, 45
   Exeter College of, 66–7, 155
   Jesus College of, 119
   Magdalen College of, 120
   Merton College of, 100
   reform of, 155
   Regius Professorship of Civil Law at, 90, 104
   Regius Professors of Modern History at, 45–6, 68, 74, 85, 92–3, 100, 116–17, 119, 154–5, 157–8, 163
   the study of history at, 55, 59, 92, 160
   Trinity College of, 45, 52, 104
   University Commission of, 82
Oxford University Press, 117

Paley, William, 49
Palgrave, Francis, 48
Pall Mall Gazette, 123
Parker, Christopher, 36
Parker, John W., 13, 23, 31
Paul, Herbert, 162
Pearson, Karl, 98
Pérez-Ramos, Antonio, 36
philology, 51, 137
phrenology, 5, 86
Pickering, Edward, 127
picturesque
   as a form of history, 113, 121–4, 139–40, 151, 160
   see also Romantic history
Poole, R. Lane, 133, 156
   as editor of the *EHR*, 141–2, 148, 152, 154
   on Acton, 147–51
   on Froude, 154

positivism, 17, 19, 21, 32
   historical scholarship and, 36–7, 41–2, 44
Porter, Theodore M., 98
Powell, Baden, 26, 31
Powell, F. York, 93, 154
popularization
   of history, 6
   of science, 5
professionalization
   of history, 2, 5–7, 9–10, 33, 36, 45, 47, 55, 58, 70, 72–3, 78–80, 82–8, 93, 97, 100, 105–6, 110–12, 119, 122, 125, 133–4, 144–5, 147, 149, 152, 159–60
   of science, 5, 7, 86, 98
   see also boundary work; objectivity; science of history
Protestantism, 3, 14, 38, 56–8, 101
   Reformation and, 22, 108
   see also Church of England; Tractarian Movement; Thirty-Nine Articles
Providence, 18, 21, 40, 44, 49, 53
Public Record Office, 45–6, 145
publishers and publishing, 7–8, 76, 100–11, 115–6
   Creighton and, 127, 130–2
   Freeman and, 116–18, 124–6
   Green and, 121–3
   Seeley and, 75–6, 128–30
   see also Macmillan, Alexander; Longman, Charles; *English Historical Review*
Pusey, Edward, 56

Quarterly Review, 124
Quetelet, Adolphe, 20–1

Rambler, 37, 39, 148
Ranke, Leopold von, 96, 115
   Acton and, 38, 97, 106, 149
   historical method of, 3–5, 37–8, 49, 68, 81, 83–5, 108
   Histories of the Latin and Germanic Nations, 38
Riehl, Wilhelm Heinrich, 149
Renouard, Pierre-Victor, 42
Revue Historique, 107
Richelieu, Cardinal, 41

Rigney, Ann, 3
Ripon Grammar School, 45
Ritter, Heinrich,
   Ancient Philosophy, 42
Rivington publishers, 127
Rolls Series, 45, 145–6
Roman Empire, 161
Romantic history, 2, 4, 5, 7–8, 74, 79–81, 98–100, 113, 147
Romantic historism, 3
   see also Romantic history
Romanticism, 106
   Buckle and, 25–6
   Mill and, 23
Romantic poetry, 24
Romilly, Samuel, 22
Roscher, Wilhelm, 149
Royal Geographical Society, 136
Royal Historical Society, 131
Royal Institution, 13–14, 51, 70, 72, 155
Ruskin, John, 160–71

Salisbury, Lord, 154
Saturday Review, 58
   as promoter of scholarly standards of history, 47, 120, 134
   on Acton, 96, 107
   on Burrows, 84–5
   on Freeman, 137, 151
   on Froude, 87, 89, 157–8, 161
   on Green, 123
   on Seeley, 77–8, 139–40
   on the *EHR*, 130
Sandars, T. C., 58
Scott, Walter, 2–3, 7, 164
   Waverly Novels, 79
   Seeley and, 79
science of history, treatment of, 7, 73–4, 93, 115, 128, 151–2, 164
   by Acton, 3, 43–4, 96–9, 106, 108–11, 147, 149
   by Bryce, 105–6
   by Buckle, 1–2, 4, 14–15, 17–23, 35–6, 164
   by Bury, 162
   by Creighton, 113–14, 147
   by Freeman, 3, 52–3, 74, 81–2, 95, 117–18, 147
   by Froude, 4, 70–2, 155–6, 159
   by Kingsley, 58–61
   by Seeley, 3, 74, 77–81, 85, 95, 129–30, 139, 147
   by Stubbs, 3, 4, 35–6, 47–51, 95, 146–7
   by Trevelyan, 163
scientific naturalism, 5
Seeley, J. R., 8, 73, 101–2, 152
   as disciplinarian, 6, 74
   as Regius Professor of Modern History at Cambridge, 74, 78, 95, 128
   concern with audience, 6, 76, 79–81, 95, 128–30
   death of, 133
   Freeman and, 150
   Gladstone and, 76–7
   Green and, 77, 130
   historical sketch of, 74–6
   inaugural lecture of, 77
   Kingsley and, 74–5
   Macmillan and, 75–6, 128–30
   on Buckle, 53, 78–9
   on Christianity and science, 75
   on Macaulay, 79–80, 85
   on Scott, 79
   on the science of history, 3, 7, 78–81, 95, 98, 139, 141, 147
   popularity of, 129, 140
   posthumous opinions of, 138–41, 151
   Saturday Review on, 77–8
   style of, 139–41, 151
   works of,
      Ecce Homo, 74–7, 128–9, 139–40
      Expansion of England, 129–30, 140
      Life and Times of Stein, 78, 139–40
      Natural Religion, 128–9, 139–40
Seeley, Mary, 129
Seeley, Robert Benton,
   Essays on the Church, 75
Seignobos, Charles,
   Introduction to the Study of History, 93
Shakespeare, William, 26, 71, 155–6
Shelley, Percy Bysshe, 66
Shirreff, Emily, 31
Sickel, Karl-Ernst, 149
Sidgwick, Henry, 148
Simpson, Richard, 37, 39–41
Smith, Adam, 24

Smith, Goldwin, 35, 45–6
    Freeman and, 92, 116–17
    Froude and, 68–71, 87
    Irish History, 44
Sobel Effect, 164
Somerset Archaeological Society, 120
Speaker, 96, 99, 135, 137, 148, 160, 162
Spectator, 91, 143, 146
Spencer, Herbert, 17
Sprengel, Matthias Christian, 42
statistics, 2, 20–2, 40–1, 61
Stephen, Leslie, 17, 33
    on Buckle, 13–15, 21
Strauss, David
    Das Leben Jesu, 75
Strachey, Lytton, 90
Stubbs, William, 63–4, 77, 83–4, 102, 132, 138
    Acton and, 45, 98, 142
    archival research of, 4, 45, 48, 55, 142
    as Bishop of Oxford, 92, 98, 111, 141–3
    as editor of the Rolls Series, 45–6, 120, 145–6
    as Regius Professor of Medieval and Modern History at Oxford, 45–7, 59, 74, 85, 92, 100, 117, 119–20, 142, 144–5, 158
    as unpopular professor, 85, 92, 158
    as whig historian, 9
    audience of, 8, 144–6
    biographical sketch of, 45
    Bryce and, 104, 152
    death of, 133, 147
    Freeman and, 35–6, 46, 48, 52–3, 90–2, 120–2, 144
    Green and, 48, 120, 122, 124, 142
    inaugural lecture of, 47–9, 144, 147, 162
    Maitland on, 141–7, 151
    on Buckle, 35–6, 47
    on controversy, 55
    on Divine Providence, 49, 53
    on Kingsley, 58, 144
    on the science of history, 3–4, 36, 47–9, 73, 97, 142, 144, 146–7
    posthumous opinions of, 142–7
    scientific racism of, 50–1
    style and art of, 144–6, 151
    works of,
        Constitutional History of England, 7, 49–51, 124, 142–6, 149
        Registrum sacrum Aglicanum, 36
sublime, 3, 24, 18, 164
Sybel, Heinrich von, 106, 149

Tanner, J. R.,
    on Seeley, 138–41, 151
Taylor, Helen, 24
Tennyson, Alfred, 152
Teutonic race,
    English inheritance of, 50–3
    political institutions of, 50
Thackeray, William Makepeace,
    The History of Henry Esmond, 3
Thirty-Nine Articles, 57, 66–7, 97
Thompson, Edith, 125–6
Todhunter, Maurice, 139–41
Tout, T. F., 154, 156
Tractarian Movement, 45, 47, 52, 68, 155
    Tracts for the Times, 56
    Remains of the Reverend R. Hurrell Froude, 65
transmutation *see* evolution
Treitschke, Heinrich von, 97, 106
Trevelyan, G. M., 163
    England in the Age of Wycliffe, 162
uniformitarian geology, 17, 57
    see also Lyell, Charles
utilitarianism, 24–5, 77
    see also Bentham, Jeremy

Vestiges of the Natural History of Creation,
    as target of boundary work, 86
    comparison with *Ecce Homo*, 76
    comparison with *History of Civilization in England*, 30, 33
    see also Chambers, Robert
Victoria, Queen, 56, 133, 157

Ward, A. W., 126
Wegele, Franz von, 102–3
Westminster Review, 32–3, 62
Whewell, William, 16
    on Buckle, 13, 29
    on induction, 25, 29

works of,
  History of the Inductive Sciences, 25
  Philosophy of the Inductive Sciences, 25
Whig, political orientation of, 87
whig interpretation of history, 9, 68
White, Hayden, 164
Wilberforce, Samuel (Bishop of Oxford), 58, 83
  Huxley and, 120
Wilberforce, William, 22
William III, 22
William the Conqueror, 118
women,
  as educators, 125
  as historians, 124–38
  science and, 26–9, 125, 127
  'nature' of, 28, 125
Woodhead, Mrs, 17
Wordsworth, William, 24